Anatomy and Physiology
A Self-Instructional Course

Anatomy and Physiology
A Self-Instructional Course

1. The Human Body and The Reproductive System
2. The Endocrine Glands and The Nervous System
3. The Locomotor System and The Special Senses
4. The Respiratory System and The Cardiovascular System
5. The Urinary System and The Digestive System

2. The Endocrine Glands and The Nervous System

Written and designed by
Cambridge Communication Limited

Medical adviser

Bryan Broom MB BS(Lond)
General Practitioner
Beit Memorial Research Fellow
Middlesex Hospital Medical Research School

SECOND EDITION

Churchill Livingstone
EDINBURGH LONDON MELBOURNE AND NEW YORK 1985

CHURCHILL LIVINGSTONE
Medical Division of Longman Group Limited

Distributed in the United States of America by Churchill Livingstone Inc., 650 Avenue of the Americas, New York, N.Y. 10011, and by associated companies, branches and representatives throughout the world.

First edition 1977
Second edition 1985
Reprinted 1991, 1993

ISBN 0-443-03206-8

British Library Cataloguing in Publication Data
A catalogue record for this book is available from the British Library

Library of Congress Cataloging in Publication Data
Anatomy and physiology.
Rev. ed. of: anatomy and physiology / Ralph Rickards, David F. Chapman. 1977.
Contents: 1. The human body and the reproductive system — 2. The endocrine glands and the nervous system — 3. The locomotor system and the special senses — [etc.]
1. Human physiology — Programmed instruction. 2. Anatomy, Human — Programmed instruction. I. Broom, Bryan. II. Rickards, Ralph. Anatomy and physiology. III. Cambridge Communication Limited.
QP34.5.A47 1984 612 84-4977

The publisher's policy is to use **paper manufactured from sustainable forests**

Printed in Hong Kong
WC/03

Contents

The Endocrine Glands 1-36
The Nervous System 37-82

Contents

The Endocrine Glands

1	INTRODUCTION	2
1.1	The endocrine glands	2
1.2	The structure of hormones	3
	TEST ONE	5
	ANSWERS TO TEST ONE	6
2	THE PITUITARY GLAND	7
2.1	Introduction	7
2.2	The anterior pituitary	8
2.3	The anterior pituitary hormones	9
2.4	The posterior pituitary	10
	TEST TWO	11
	ANSWERS TO TEST TWO	12
3	THE SUPRARENAL GLANDS	13
3.1	Introduction	13
3.2	The hormones of the suprarenal cortex	14
3.3	The mineralocorticoids	15
3.4	The glucocorticoids	16
3.5	The production of steroids	17
3.6	Disorders of corticosteroid production	18
3.7	The hormones of the suprarenal medulla	19
	TEST THREE	21
	ANSWERS TO TEST THREE	22
4	THE THYROID, PARATHYROID, PANCREAS AND THYMUS	24
4.1	The thyroid gland	24
4.2	Disorders of the thyroid gland	25
4.3	The parathyroid glands	26
4.4	Disorders of the parathyroid glands	27
4.5	The pancreas	28
4.6	Diabetes mellitus	29
4.7	The thymus gland	30
	TEST FOUR	31
	ANSWERS TO TEST FOUR	32
	POST TEST	33
	ANSWERS TO POST TEST	35

1. Introduction

1.1. The endocrine glands

A *gland* is an organ whose main function is to produce a biologically useful substance. Often this *secretion* passes down a **duct** to its site of action, as for example in sweat, gastric or salivary glands.

However, certain glands — the endocrine glands — have no duct. The secretions, or *hormones*, from these glands pass directly into the **blood stream** and can have a widespread effect.

The hormones act as chemical messengers and provide a way in which the body can co-ordinate its functions.

Many of the endocrine glands are under the control of the nervous system via the **hypothalamus** of the brain, and the **pituitary gland** which is suspended from it.

The other principal endocrine glands are:

the **thyroid** and **parathyroid** glands

the **suprarenal**, or **adrenal** glands;

the **islets of the pancreas** (islets of Langerhans);

the **testes** (in the male)
or **ovaries** (in the famale).

Some organs produce hormones which have a purely local effect. Thus the stomach and upper digestive tract produce hormones which, via the blood stream, control the secretion of digestive glands. The kidneys produce renin which affects blood pressure, and erythropoietin which affects red blood cell production. These hormones are considered with their systems.

1.2. The structure of hormones

Hormones are of various chemical types.

They may be *proteins* (long complex folded chains of amino acids)
e.g. the pancreatic hormone insulin;

or *peptides* (simpler short chains of amino acids)
e.g. the posterior pituitary hormones;

or *glycoproteins* (complexes of proteins and carbohydrates)
e.g. thyroid stimulating hormone;

or *simple aromatic compounds*
e.g. thyroxine;

or *steroids*
e.g. the hormones of the adrenal cortex and the sex hormones.

These compounds are only present in minute amounts. They probably exert their effects on the cells of their target tissues in a variety of ways.

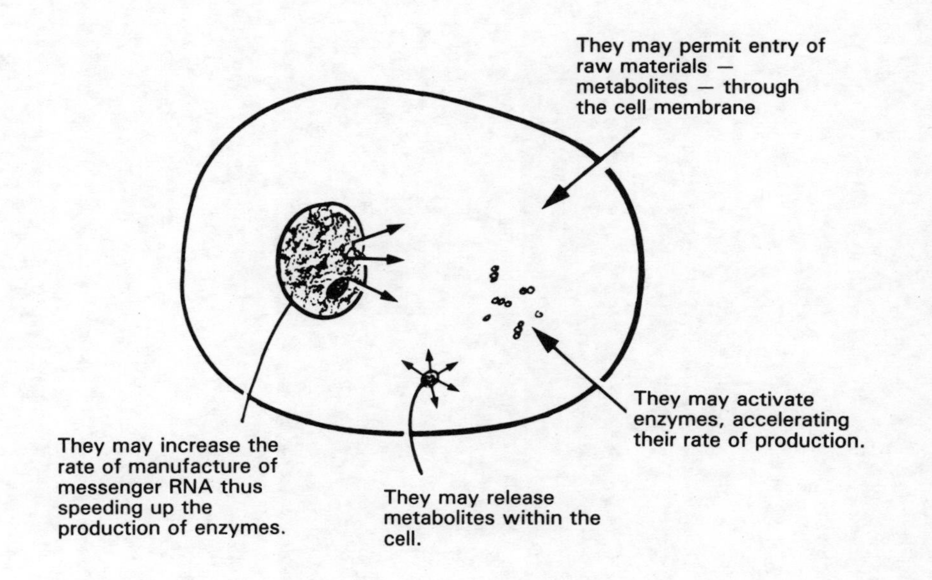

TEST ONE

1. How do endocrine glands differ from other kinds of glands?

____

____

2. Which of the following types of chemical occur as hormones?

(a) Proteins.
(b) Lipids.
(c) Steroids.
(d) Carbohydrates.
(e) Peptides.

3. Name four endocrine glands.

____

____

____

____

ANSWERS TO TEST ONE

1. Unlike other types of glands endocrine glands are ductless; their secretions or hormones pass directly into the bloodstream and so can have far reaching effects.

2. (a) Proteins.
(c) Steroids.
(e) Peptides.

3. Four of the following:

The hypothalamus.
The pituitary gland.
The thyroid gland.
The parathyroid gland.
The suprarenal glands.
The islets of the pancreas.
The testes.
The ovaries.

2. The pituitary gland

2.1. Introduction

The **pituitary gland** (or *hypophysis cerebri*) controls thyroid and sexual function, growth, and the metabolism of water, protein, fat and carbohydrate.

It is about 1 cm in diameter and occupies a hollow in the sphenoid bone called the *sella turcica.* This bony pit lies in the base of the skull, at the back of the nose, above the sphenoid air sinus.

It is suspended from the **hypothalamus,** the mass of nervous tissue which forms the floor of the third ventricle.

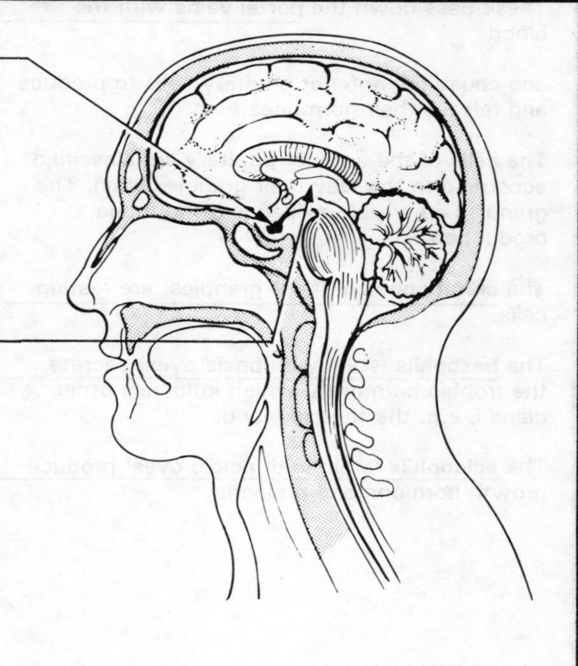

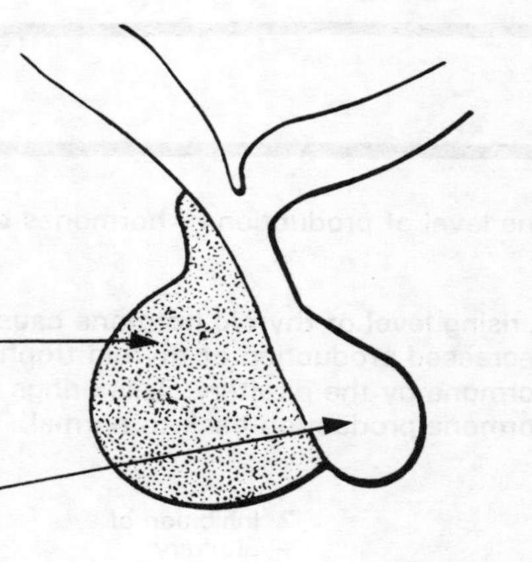

In man the gland has two major parts having different origin and functions;

the **anterior pituitary** (which controls other endocrine glands is derived from an up-growth of the lining of the primitive pharynx in the embryo);

the **posterior pituitary** (is a down-growth from the forebrain).

Both parts are under the control of the hypothalamus.

Capillaries supplying blood to the hypothalamus run together to form several **portal veins** which supply the anterior lobe. These veins carry hormones from the hypothalamus to the anterior pituitary.

Nerve fibres from the hypothalamus run to the **posterior pituitary.**

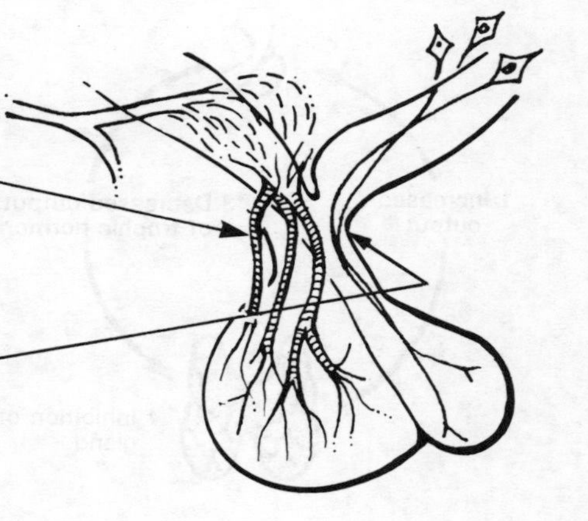

2.2. The anterior pituitary

Nerve centres in the hypothalamus produce releasing hormones.

These pass down the **portal veins** with the blood

and cause the anterior pituitary cells to produce and release their hormones.

The cells of the anterior pituitary are classified according to the way their granules stain. The granules are visible evidence of hormone production.

The **chromophobes** (faint granules) are resting cells.

The **basophils** (stain with basic dyes) secrete the trophic hormones, which influence other glands, e.g. the thyroid gland.

The **acidophils** (stain with acidic dyes) produce growth hormone and prolactin.

The level of production of hormones by a target gland is controlled by *negative feedback*.

A rising level of thyroid hormone causes a decreased production of thyroid trophic hormone by the pituitary. This brings thyroid hormone production back to normal.

A falling level of thyroid hormone stimulates the production of thyroid trophic hormone by the pituitary. This increases the output of thyroid hormone, restoring the normal level.

2 Inhibition of
— pituitary
1 Increased output
3 Decreased output of trophic hormone
4 Inhibition of gland

2 Stimulation
+ of pituitary
1 Decreased output
3 Increased output of trophic hormone
4 Stimulation of gland

2.3. The anterior pituitary hormones

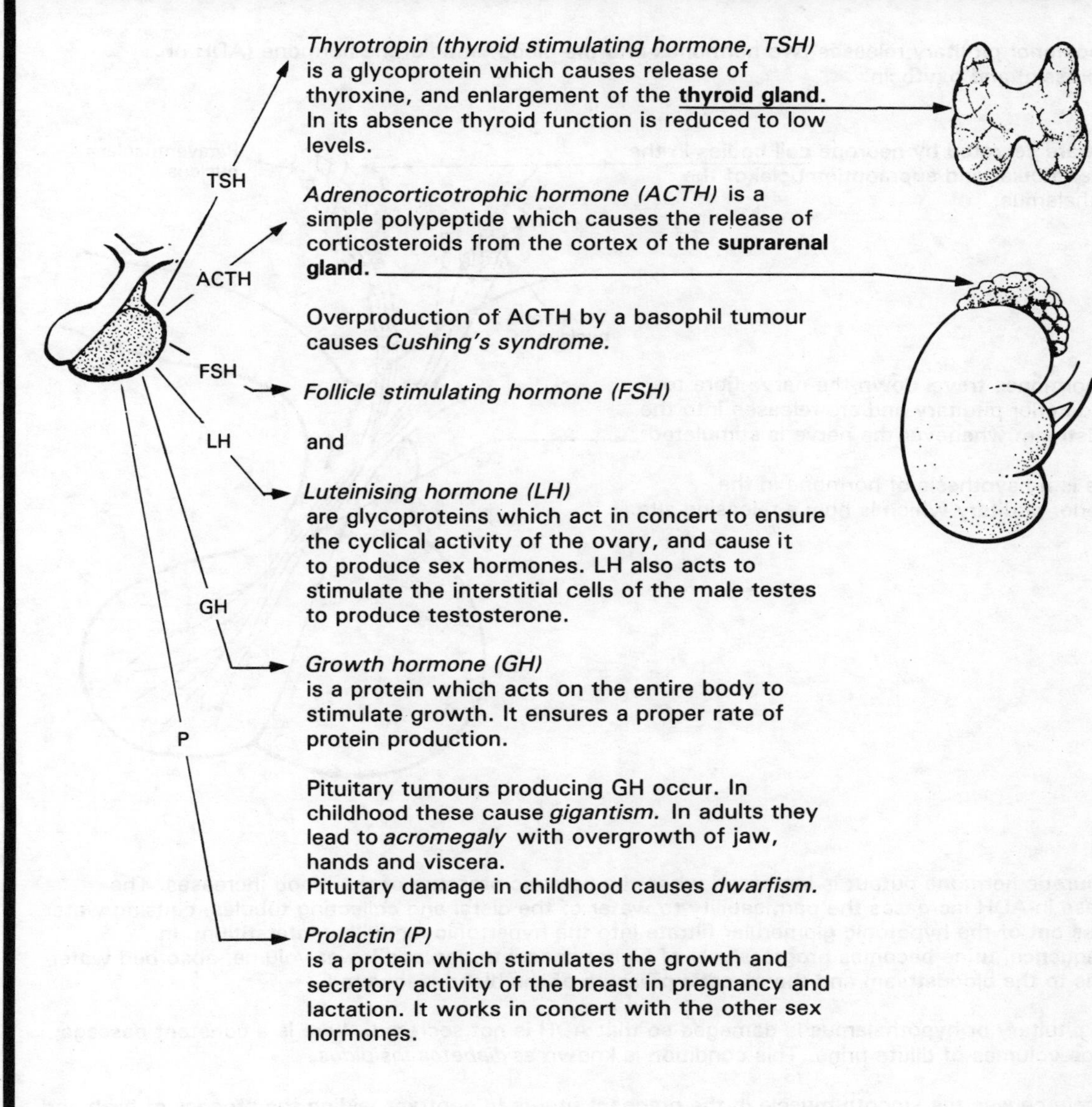

Thyrotropin (thyroid stimulating hormone, TSH) is a glycoprotein which causes release of thyroxine, and enlargement of the **thyroid gland.** In its absence thyroid function is reduced to low levels.

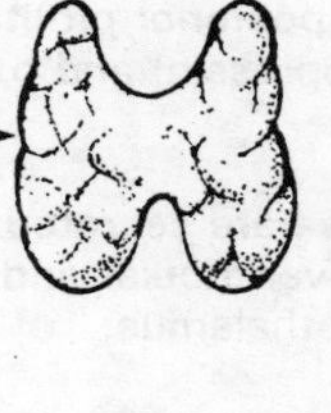

Adrenocorticotrophic hormone (ACTH) is a simple polypeptide which causes the release of corticosteroids from the cortex of the **suprarenal gland.**

Overproduction of ACTH by a basophil tumour causes *Cushing's syndrome.*

Follicle stimulating hormone (FSH)

and

Luteinising hormone (LH)
are glycoproteins which act in concert to ensure the cyclical activity of the ovary, and cause it to produce sex hormones. LH also acts to stimulate the interstitial cells of the male testes to produce testosterone.

Growth hormone (GH)
is a protein which acts on the entire body to stimulate growth. It ensures a proper rate of protein production.

Pituitary tumours producing GH occur. In childhood these cause *gigantism.* In adults they lead to *acromegaly* with overgrowth of jaw, hands and viscera.
Pituitary damage in childhood causes *dwarfism.*

Prolactin (P)
is a protein which stimulates the growth and secretory activity of the breast in pregnancy and lactation. It works in concert with the other sex hormones.

Destruction of the anterior pituitary (by tumour or ruptured blood vessels) causes *hypopituitarism,* with slowing of metabolism, extreme sensitivity to stress, weakness, skin pallor, cessation of menstruation, and loss of secondary sexual characteristics. Such individuals are very liable to die from minor injuries or trivial fasting.

2.4. The posterior pituitary

The posterior pituitary releases two hormones into the blood, antidiuretic hormone (ADH or vasopressin) and oxytocin.

These are secreted by neurone **cell bodies** in the paraventricular and supraoptic nuclei of the hypothalamus.

Paraventricular nucleus

Supraoptic nucleus

The hormones travel down the nerve fibre to the posterior pituitary and are released into the bloodstream whenever the nerve is stimulated.

There is no synthesis of hormone in the posterior pituitary which is only a releasing site.

Antidiuretic hormone output is increased when the osmotic pressure of the blood increases. The increase in ADH increases the permeability to water of the distal and collecting tubules, causing water to pass out of the hypotonic glomerular filtrate into the hypertonic medullary interstitium. In consequence, urine becomes progressively of higher concentration and lower volume, absorbed water returns to the bloodstream and the osmotic pressure of the blood is lowered.

If the pituitary or hypothalamus is damaged so that ADH is not secreted, there is a constant passage of large volumes of dilute urine. This condition is known as *diabetes insipidus.*

Oxytocin causes the smooth muscle in the pregnant uterus to contract, aiding the process of birth and helping the uterus to return to a normal size after birth. It also causes release of milk from the lactating breast by causing the myoepithelial cells to contract. Suckling causes reflex release of oxytocin by stimulation of the nipple.

TEST TWO

1. (a) Label the lettered parts of the diagram alongside.

(b) Which of these features contain secretory cells?

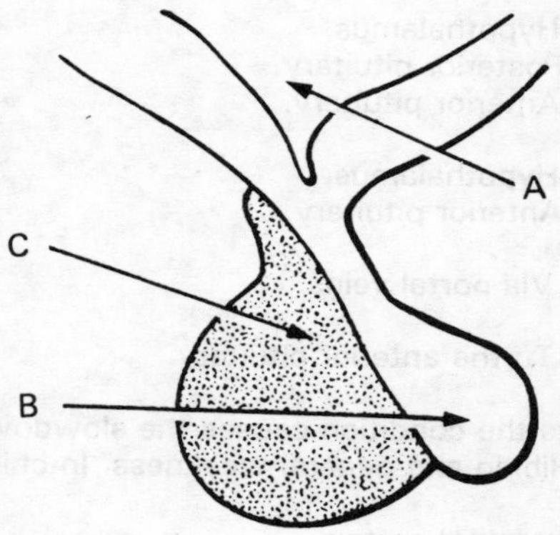

2. **(i) How do releasing hormones travel from the hypothalamus?**
(a) Via the general circulation. (b) Via portal veins. (c) Via nerve fibres.

(ii) To where do these hormones travel?
(a) To the anterior pituitary. (b) To the posterior pituitary. (c) To the thyroid gland.

3. What is hypopituitarism, and what are the effects of this condition?

4. Which of the following endocrine glands are controlled by the pituitary gland?

(a) Suprarenal cortex.

(b) Pancreas.

(c) Ovary.

(d) Parathyroid.

5. Which of the statements on the right apply to the hormones listed on the left?

(i)	Adrenocorticotrophic hormone.	(a)	Responsible for maintaining the osmotic pressure of the blood.
(ii)	Luteinising hormone.	(b)	Stimulates the growth of the breasts during pregnancy.
(iii)	Prolactin.	(c)	Causes the release of corticosteroids from the suprarenal cortex.
(iv)	Antidiuretic hormone.	(d)	Stimulates the testes to produce testosterone.

ANSWERS TO TEST TWO

1. (a) A. Hypothalamus.
   B. Posterior pituitary.
   C. Anterior pituitary.

   (b) A. Hypothalamus.
   B. Anterior pituitary.

2. (i) (b) Via portal veins.

   (ii) (a) To the anterior pituitary.

3. In adults the condition causes the slowdown of metabolism, pallor, the cessation of menstruation, loss of libido and general weakness. In children hypopituitarism invariably causes dwarfism.

4. (a) Suprarenal cortex.

   (c) Ovary.

5. (i) Adrenocorticotrophic hormone — (c) causes the release of corticosteroids from the suprarenal cortex.

   (ii) Luteinising hormone — (d) stimulates the testes to produce testosterone.

   (iii) Prolactin — (b) stimulates the growth of the breasts during pregnancy.

   (iv) Antidiuretic hormone — (a) is responsible for maintaining the osmotic pressure of the blood.

3. The suprarenal glands

3.1. Introduction

The **suprarenal glands** (or adrenal glands) lie on the upper poles of each kidney, just outside the renal fascia. They are golden yellow in colour and have a rich blood supply.

Each suprarenal gland is made up of two functionally different types of endocrine tissue:

the **cortex**

the **medulla**.

The **suprerenal cortex** is derived from mesoderm tissue. It is essential for life.

Three separate zones of tissue can be identified in the cortex:

— the **zona glomerulosa** made up of clumps of small cells which secrete mineralocorticoids;

— the **zona fasciculata** (the main mass) made up of columns of cells which secrete the glucocorticoids (and some sex hormones);

— the **zona reticularis,** an irregular network of resting cells which can be called on in an emergency.

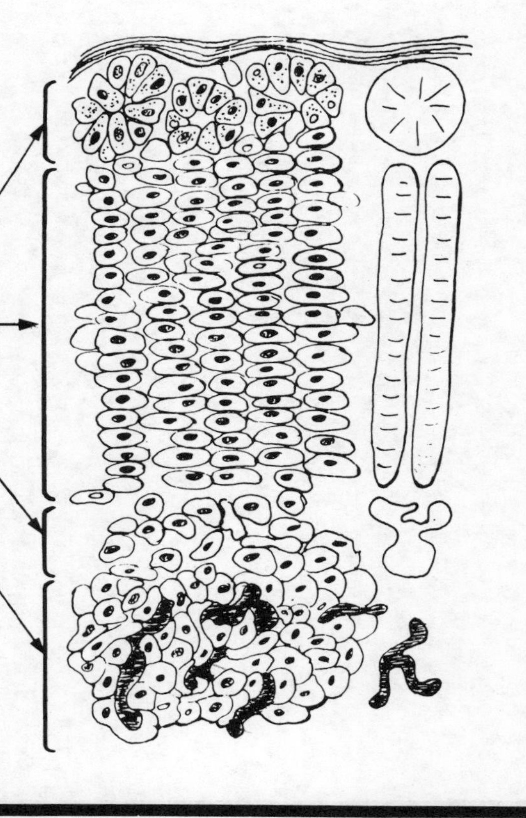

The **suprarenal medulla** is composed of a mass of small *chromaffin* cells with venous sinuses between them.

It is derived from primitive nervous tissue, and is functionally related to the autonomic nervous system — it secretes adrenalin and noradrenalin. The suprarenal medulla is not essential to life.

3.2. The hormones of the suprarenal cortex

Over 40 different steroids have been isolated from the suprarenal cortex, but only a few of these have been detected in the venous blood leaving the suprarenal gland. The remainder must represent stages in hormone synthesis. They are all derived from cholesterol, a steroid widespread in nature.

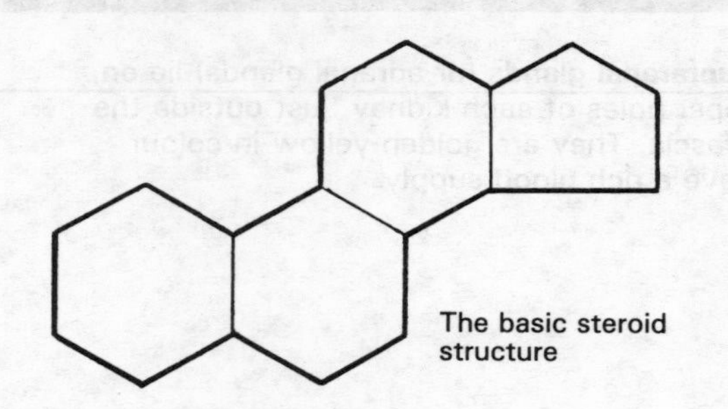
The basic steroid structure

The steroid hormones fall into three main groups:

1. *the mineralocorticoids* — of which the most important is *aldosterone;*

2. *the glucocorticoids* — of which *cortisol* (or hydrocortisone) is the most imporant;

3. *the sex hormones* — these are androgens and oestrogens. They are of no great importance in the normal subject since their main sites of production are the ovaries or testes.

3.3. The mineralocorticoids

Aldosterone, and other similar compounds such as deoxycorticosterone, alter the permeability of cell membranes to electrolytes especially to sodium ions and potassium ions.

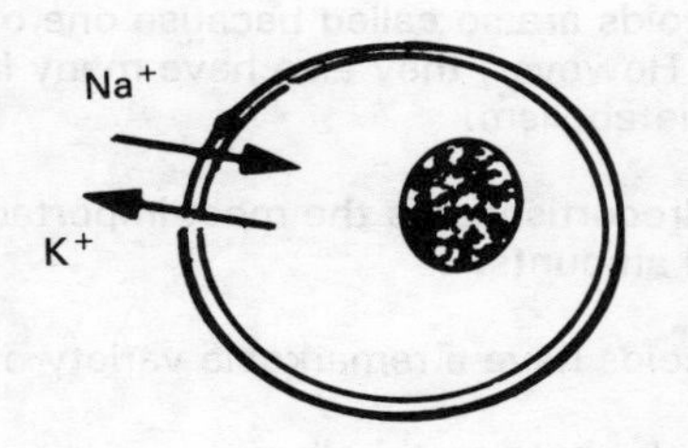

The most important action of aldosterone is on the tubules of the kidney. An increase in aldosterone secretion increases the reabsorption of sodium back into the blood from the urine and hence conserves sodium within the body. Simultaneously it increases the excretion of potassium from the blood into the urine, and thus an electrolyte balance is maintained.

Na^+

K^+

A similar, but less important, effect occurs in sweat and digestive secretions.

Lack of aldosterone in the body causes sodium and water loss with a consequent decrease in blood volume, circulatory collapse with low blood pressure, and ultimately, death.

3.4. The glucocorticoids

The glucocorticoids are so called because one of their main effects is on the metabolism of carbohydrates. However, they also have many important properties which are not associated with carbohydrate metabolism.

Cortisol (or hydrocortisone) is the most important glucocorticoid. Corticosterone and cortisone are also formed in small amounts.

The glucocorticoids have a remarkable variety of effects.

Effects on carbohydrate metabolism
They increase glucose synthesis from non-carbohydrate sources, such as amino acids (gluconeogenesis).

They increase the storage of glucose as glycogen within the liver.

They decrease the use of glucose by the body tissues.

The net result is to raise the blood sugar level.

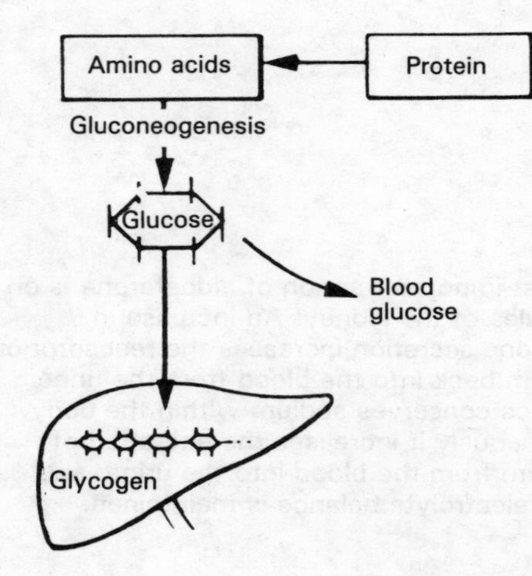

Effect on protein metabolism
They decrease protein synthesis throughout the body, since the amino acids are diverted to gluconeogenesis. However, in the liver, protein synthesis increases.

The net result is to cause loss of tissue proteins and an increased output of nitrogen (as urea) in the urine.

Effect on fat metabolism
They mobilise fatty acids from fat stores in adipose tissue, leading to an increase in the blood of fatty acids which can be used as an energy source by the tissues.

Effects on the blood
They increase the production of red blood cells.

They decrease the production of eosinophils.

Other effects
They stabilise lysozymes within cells.

They have a weak mineralocorticoid action, retaining sodium.

They maintain the blood pressure, by an action on the blood vessels and the heart.

They maintain the normal activity of voluntary muscle, which becomes weak in its absence.

In large amounts (not seen in nature) the glucocorticoids have an anti-inflammatory and anti-allergic effect, reducing the extent of oedema, blood vessel dilation, white blood cell invasion and other effects occurring in the inflammatory reaction to an injury. Output of these hormones is increased about six times in response to stresses such as anxiety and injury.

3.5. The production of steroids

The production of aldosterone depends on the presence in the blood stream of a hormone called angiotensin II.

Angiotensin II is formed by the action of the enzyme renin on a plasma globulin.

Renin is released from the kidney in response to sodium depletion, potassium overload or a fall in blood volume. Aldosterone tends to counteract these conditions resulting, for example, from vomiting, dehydration or injury.

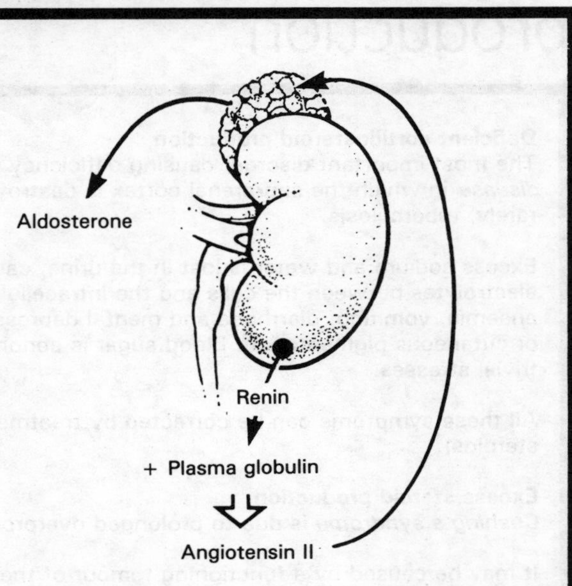

The secretion of cortisol depends on the secretion of ACTH by the pituitary gland. If ACTH secretion stops, cortisol secretion is reduced to very low levels. If ACTH secretion is suppressed for long periods the suprarenal cortex thins and may waste away. Secretion of ACTH depends in its turn on the secretion of corticotrophin release hormone by the hypothalamus.

There is a regular 24 hour (circadian) variation in cortisol output which reflects the rhythmical activity of the hypothalamus. (There is a similar diurnal variation in body temperature which is also under hypothalamic control.)

Circulating corticosteroids, above a certain level, inhibit the production of corticotrophin releasing hormone, and hence the secretion of ACTH. Corticosteroid production is promptly reduced.

This negative feedback mechanism tends to maintain cortisol at a stable level.

Physical stress or prolonged anxiety acts via nervous influences on the hypothalamus to cause an increase in cortisol levels. Without this increase a person has little resistance to injury and is liable to die during a stressful experience such as a minor operation.

Steroids given for the treatment of disease suppress ACTH, just as natural cortisol does, and if continued for a long time may lead to atrophy of the cortex. The patient is then dependent on steroid therapy for dealing with stressful experiences.

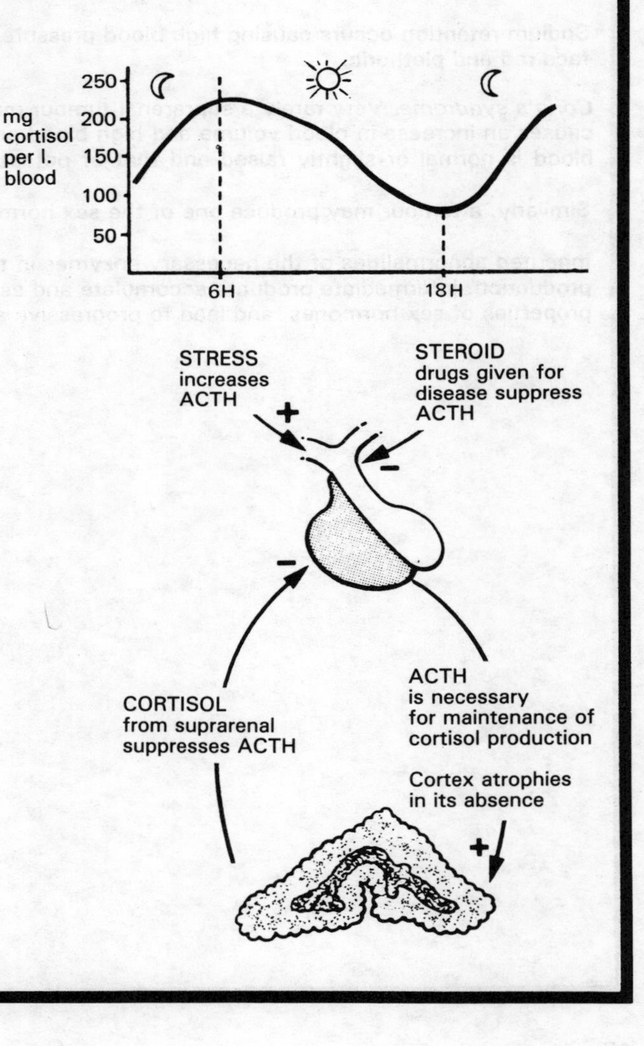

3.6. Disorders of corticosteroid production

Deficient corticosteroid production
The most important disorder causing deficiency in the production of corticosteroids is *Addison's disease* in which the suprarenal cortex is destroyed by auto-immune disease, secondary carcinoma or rarely, tuberculosis.

Excess sodium and water is lost in the urine, causing a low blood pressure. The distribution of electrolytes between the cells and the intracellular fluid is disturbed and results in muscle weakness, anaemia, vomiting, diarrhoea and mental depression. The characteristic physical sign is some mucous or cutaneous pigmentation. Blood sugar is abnormally low and there is an inability to tolerate even trivial stresses.

All these symptoms can be corrected by treatment with cortisol and aldosterone (or synthetic steroids).

Excess steroid production
Cushing's syndrome is due to prolonged overproduction of **cortisol.**

It may be caused by a functioning tumour of the suprarenal cortex, or a basophil tumour of the pituitary which produces excess ACTH. Overdosage of synthetic steroids in the treatment of certain diseases may produce a similar condition.

The patient has a fat trunk and 'moon face', with thin limbs, due to abnormalities of fat deposition. Abnormal protein breakdown causes muscle wasting, thinning of the skin dermis with broad stretch marks, and loss of bone collagen leading to spontaneous fractures. The blood sugar is high, and sugar is secreted in the urine (diabetes mellitus).

Sodium retention occurs causing high blood pressure, and overproduction of red blood cells makes the face red and plethoric.

Conn's syndrome. Very rarely a suprarenal tumour may produce an excess of **aldosterone,** which causes an increase in blood volume and high blood pressure. The concentration of sodium ions in the blood is normal or slightly raised and that of potassium is low.

Similarly, a tumour may produce one of the sex hormones, causing abnormal virilism or feminisation.

Inherited abnormalities of the necessary enzymes in the suprarenal cortex may block cortisol production. Intermediate products accumulate and escape into the blood stream, where they have the properties of sex hormones, and lead to progressive abnormalities of sexual development.

3.7. The hormones of the suprarenal medulla

Stressful situations (such as exercise, cold, injury or suffocation or situations causing fear, anxiety, pain, fall in blood pressure or blood glucose), cause a flood of nervous impulses to the **hypothalamus.**

Nervous impulses travel to the suprarenal medulla via the **sympathetic nerves** which derive from the thoracic spinal nerves. The suprarenal medulla is the only endocrine gland which has a rich nerve supply.

Nervous stimulation of the suprarenal medulla causes the hormones *adrenalin* and *noradrenalin* to be released into the blood stream. These hormones and their breakdown products are known as *catecholamines.* They act on tissues throughout the body. Their overall effect is to prepare the body for activity and exertion in response to threat by diverting the blood to the limb muscles, increasing the action of the heart, dilating the bronchial tree, mobilising liver glycogen to raise blood glucose, and reducing most other activity, for example, in the gut.

This *'fight-or-flight'* reaction acts to reinforce the effect of the sympathetic nervous system. It is not essential to life.

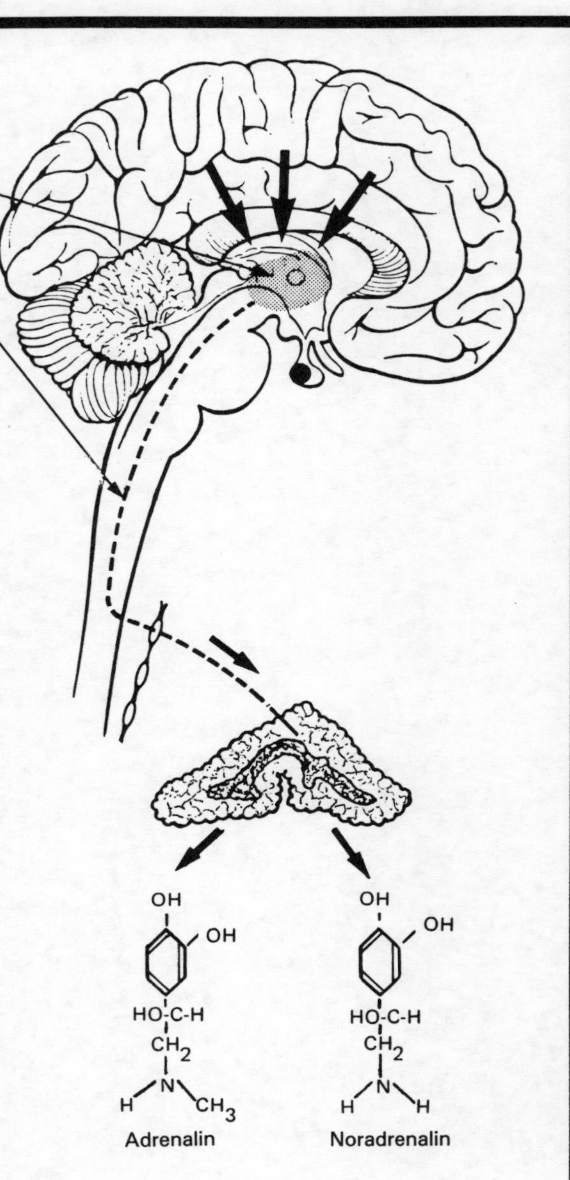

The receptors for the catecholamines are areas on the cell walls of the target tissues, which take up the hormones and transmit their effects to the cell.

There are two types of receptor for catecholamines, α-receptors and β-receptors. In general, α-receptors mediate excitatory responses (e.g. contraction of smooth muscle of gut, vasoconstriction) while β-receptors mediate inhibitory responses (e.g. dilatation of bronchial muscle). There is one major exception. The β-receptors of the heart serve to stimulate the heart.

Adrenalin and noradrenalin have different affinities for the two types of receptor and therefore give rise to different effects. Thus noradrenalin mainly causes blood vessel constriction whereas adrenalin produces a more complex reaction.

The receptors can be selectively occupied and blocked by different drugs — the α- and β-blocking drugs.

5.2 The hormones of the suprarenal medulla

TEST THREE

1. (a) **Which of the zones labelled on the diagram of the suprarenal gland alongside is essential to life?** A

   (b) **Which is the cortex?**

   A

2. **Which of the descriptions on the right apply to the hormones listed on the left?**

(i) Androgen. c	(a) A mineralocorticoid.
(ii) Aldosterone. a	(b) A glucocorticoid.
(iii) Cortisol. b	(c) A sex hormone.

3. **Among the following conditions are two that are connected with aldosterone: which are they, and what is the connection?** Lack of aldosterone will increase sodium loss + water : fall in bld vol

   Dizziness, sodium depletion, depigmentation, nausea, a fall in blood volume, gastrointestinal discomfort, potassium depletion, drowsiness, potassium overload.

4. **Which part of the body is responsible for the circadian variation in cortisol production?**

   Hypothalamus

5. **Place ticks in the appropriate brackets to indicate whether cortisol increases or decreases the following.**

		Increased by cortisol	Decreased by cortisol
(a)	Glucose production from amino acids.	(✓)	( )
(b)	Use of glucose by the tissues.	( )	(✓)
(c)	Synthesis of protein.	(✓)	( )
(d)	Red cell production.	(✓)	( )
(e)	Fatty acids in the blood.	(✓)	( )

6. (a) **What are the hormones released from the suprarenal medulla at times of stress?**

   (b) **Summarise the effect of those hormones.**

   Adrenalin Noradrenalin

   Prepare body for activity + exertion in response to threat by diverting blood to limb muscles, increasing action of heart dilating bronchial tree raise blood glucose

ANSWERS TO TEST THREE

1. (a) A

   (b) A

2. (i) Androgen is (c) a sex hormone.

   (ii) Aldosterone is (a) a mineralocorticoid.

   (iii) Cortisol is (b) a glucocorticoid.

3. A fall in blood volume and potassium overload are connected with aldosterone, which tends to counteract both conditions.

4. The hypothalamus.

5.

		Increased by cortisol	Decreased by cortisol
(a)	Glucose production from amino acids.	(√)	( )
(b)	Use of glucose by the tissues.	( )	(√)
(c)	Synthesis of protein.	( )	(√)
(d)	Red cell production.	(√)	( )
(e)	Fatty acids in the blood.	(√)	( )

6. (a) Adrenalin and noradrenalin.

   (b) They prepare the body for 'fight-or-flight'. They reinforce the effects of the sympathetic nervous system.

4. The thyroid, parathyroid, pancreas and thymus

4.1. The thyroid gland

The thyroid gland is wrapped around the front of the upper trachea. It consists of two **lobes,** joined by an **isthmus.** It receives a rich blood supply from the **superior and inferior thyroid arteries.**

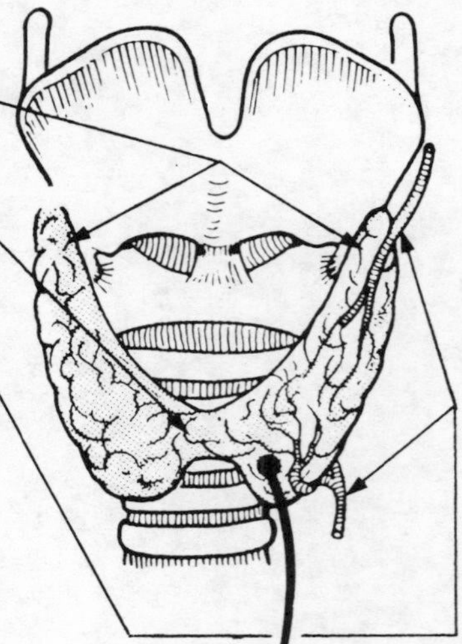

The thyroid is composed of a mass of hollow spheres or *follicles.* Each follicle has a **wall** one cell thick, and contains a jelly-like **colloid.**

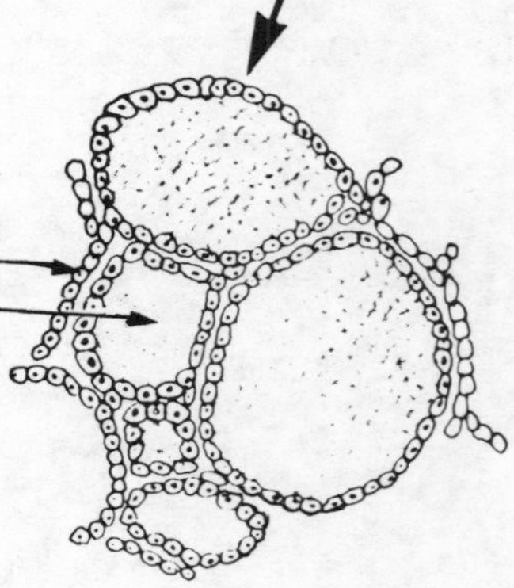

The lining cells of the follicles have a remarkable ability to extract iodine from the blood, and combine it with the amino acid tyrosine, to form an active hormone tri-iodothyronine (T3). Some less active thryoxine is also formed. Thyroxine (T4) is converted to tri-iodothyronine in the body. These compounds, and certain intermediates, are stored in the colloid of the follicles. Storage is necessary as iodine may be absent from the diet for long periods. Where this is the case, the thyroid gland becomes greatly enlarged — *goitre.* Today, lack of iodine due to the inaccessibility of sea fish, or to eating vegetables deficient in iodine owing to iodine deficiency in soil, is readily compensated by the use of genuine sea salt or table salt to which iodine has been added.

Release of T3 and thyroxine is largely under the control of TSH released from the pituitary. TSH production is inhibited by high levels of thyroid hormones.

The thyroid hormones increase the metabolic rate of all tissues, probably by increasing the synthesis of respiratory enzymes within the cell.

Calcitonin is a comparatively recently discovered polypeptide hormone which is produced by cells between the follicles. It can lower the concentration of calcium ions in the blood and cause deposition of calcium in bone. Its role is at present unknown. It may help in treating Paget's disease of the bone.

4.2. Disorders of the thyroid gland

Failure of thyroid secretion
In children failure of thyroid secretion may occur, because of a congenital absence of the necessary enzymes in the thyroid cells. It results in *cretinism.* The child is dwarfish and mentally retarded, with a thick skin, scanty hair, hoarse voice and a large protruding tongue. Early thyroxine therapy prevents the condition.

In adults the thyroid can be destroyed slowly by auto-immune disease. This results in *myxoedema* with slowing of all bodily functions, mental dullness, subnormal temperature, thick coarse skin and slow pulse.

Excess thyroid secretion
Hyperthyroidism occurs because of the abnormal synthesis within the body of compounds which act like TSH. It results in an increased metabolic activity with increased appetite and heat production. Symptoms include anxiety and excitability, a fine tremor of the hands, an intolerance of warmth, weight loss, diarrhoea, sweating and a staring expression.

4.3. The parathyroid glands

There are four **parathyroid glands** which are normally found on the posterior surface of the thyroid gland.

Each gland consists of a closely packed mass of two types of cell. One of these types of cell secretes the hormone *parathormone*.

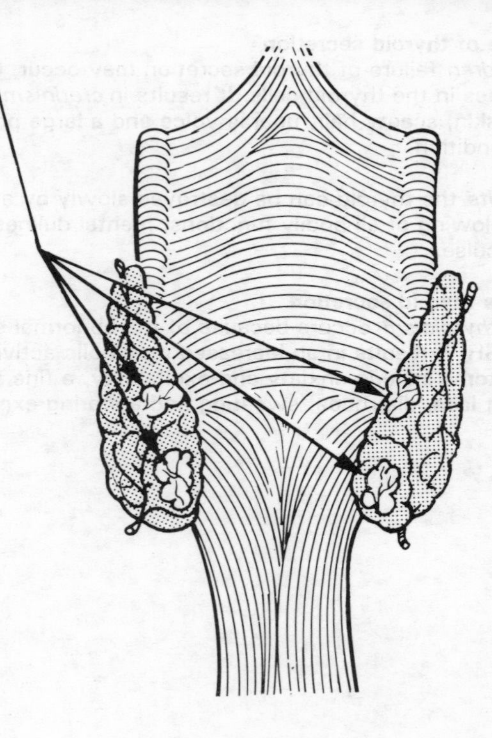

Parathormone is a polypeptide involved in the control of the calcium ion concentration of the blood. It is secreted in response to a fall in calcium ion concentration and tends to raise it.

Parathormone produces its effect by:

(i) mobilisation of calcium ions from the bones, by stimulating the bone cells;

(ii) prevention of loss of calcium ions in the urine, by promoting reabsorption of calcium ions from the tubules back into the blood;

(iii) increasing the absorption of calcium from the gut. It is however much less important than vitamin D in this respect.

Parathormone may well work in conjunction with calcitonin (which lowers the blood calcium) to keep calcium ion concentration at a stable level in the blood.

4.4. Disorders of the parathyroid glands

Deficient secretion

Deficient secretion is usually a consequence of surgery on the thyroid gland, which damages or removes the parathyroids. It results in a fall in blood calcium. After several days the patient notices 'pins-and-needles', muscular twitching, and muscular spasms. If all the parathyroid tissue has gone the condition may progress to generalised convulsions and death (which shows just how important a stable calcium level is for muscle metabolism). It may be corrected by the intravenous administration of calcium ions.

Excess secretion

Excess secretion is most commonly caused by a parathyroid tumour. It leads to a grossly raised blood calcium. The bones become decalcified and brittle, so that they may fracture spontaneously. Calcium may be deposited in the kidney causing renal failure, and in the urinary tract as stones. There may also be gross mental disturbances.

4.5. The pancreas

The bulk of the **pancreas** consists of **glandular acini** which secrete a powerful digestive juice into the gut.

Scattered amongst the acini are tiny clumps of cells known as the *islets of Langerhans.*

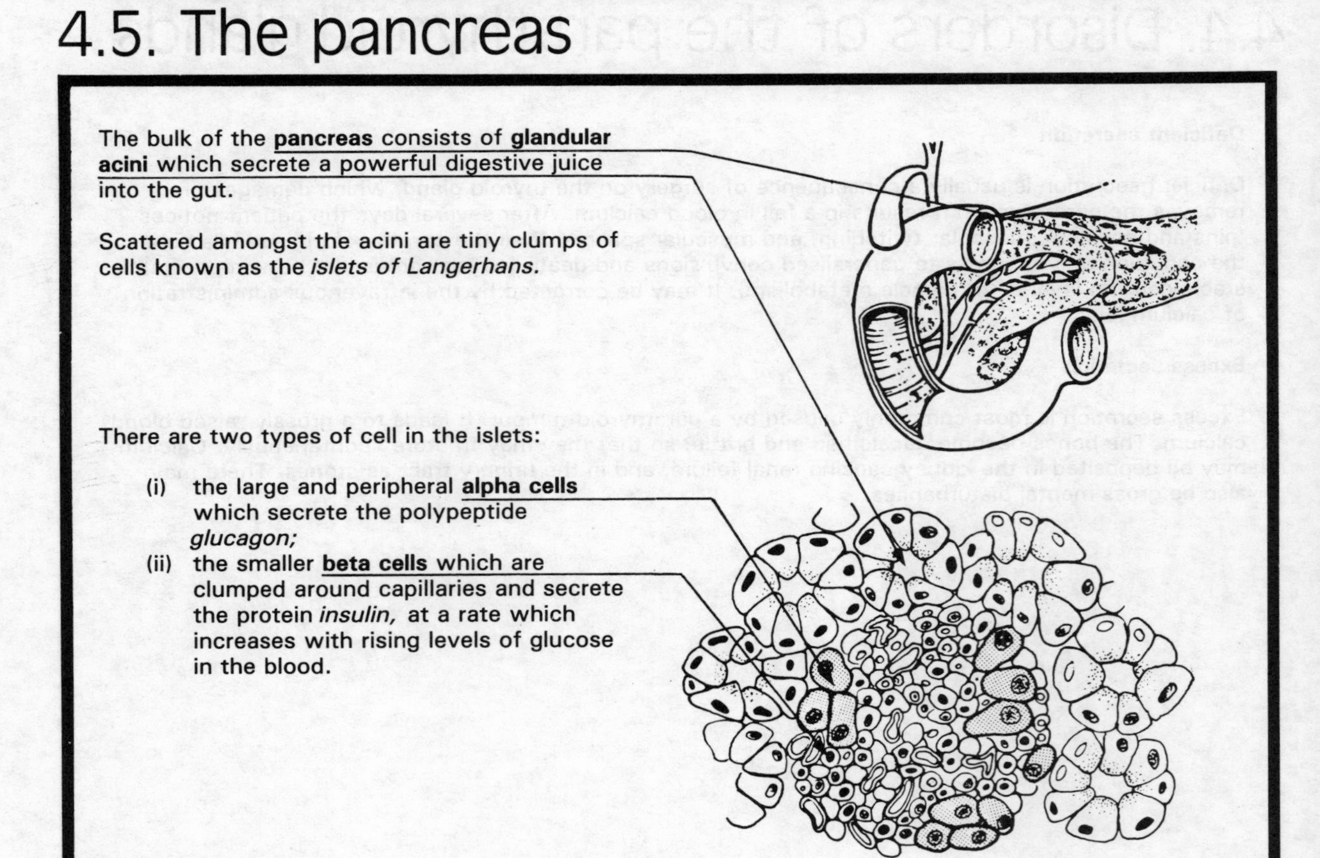

There are two types of cell in the islets:

(i) the large and peripheral **alpha cells** which secrete the polypeptide *glucagon;*

(ii) the smaller **beta cells** which are clumped around capillaries and secrete the protein *insulin,* at a rate which increases with rising levels of glucose in the blood.

Glucagon increases blood sugar levels by mobilising glycogen from the liver. It also mobilises stored fat and causes the rapid release of insulin from the islets.

Insulin is essential to life. Its fundamental effect is to increase the transport of glucose into cells. It also increases protein synthesis in all cells, increases the laying down of fat in adipose tissue, and is necessary for the full breakdown of fatty acids by the liver.

Insulin lowers blood sugar levels, partly because it encourages the 'mopping up' of glucose by the body's cells, and partly because it stops the production of glucose from amino acids in the liver.

4.6. Diabetes mellitus

If inslulin secretion is reduced or stops, due to loss of beta cells, the level of sugar in the blood becomes very high. The kidney is unable to reabsorb all the glucose and some of it escapes in the urine accompanied by water and electrolytes. Furthermore, body protein breaks down, and partly metabolised fat products, called ketones, accumulate. This condition is known as *diabetes mellitus.*

The patient with diabetes mellitus has a high output of sugar-laden urine, thirst because of water loss, wasting and weight loss. Ultimately coma may occur because of the accumulation of waste products. The condition is treated by the administration of insulin.

Glucagon, also a pancreatic hormone, may be useful in the treatment of hypoglycaemia. It is an insulin antagonist.

Growth hormone and *cortisol* are both insulin antagonists. They reduce the uptake of glucose by cells and so raise the blood sugar.

Fatty acids in high concentrations have a similar effect.

In patients with acromegaly and Cushing's syndrome, and in some overweight subjects, carbohydrate metabolism is impaired.

4.7. The thymus gland

The thymus lies behind the sternum, in front of the lungs and heart. It is of vital importance in the development of the lymphatic system.

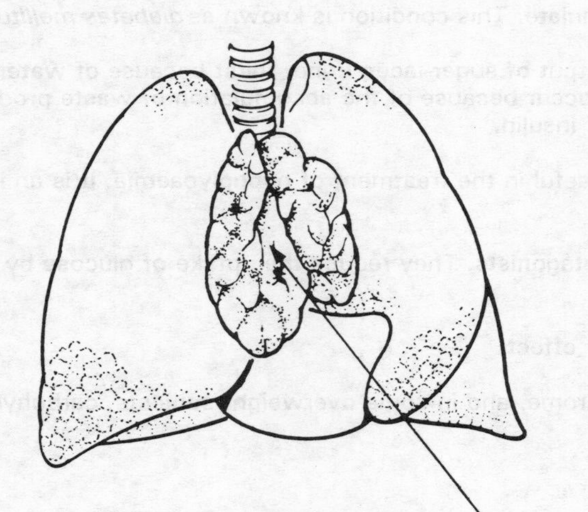

The thymus is relatively large and fleshy in infancy.

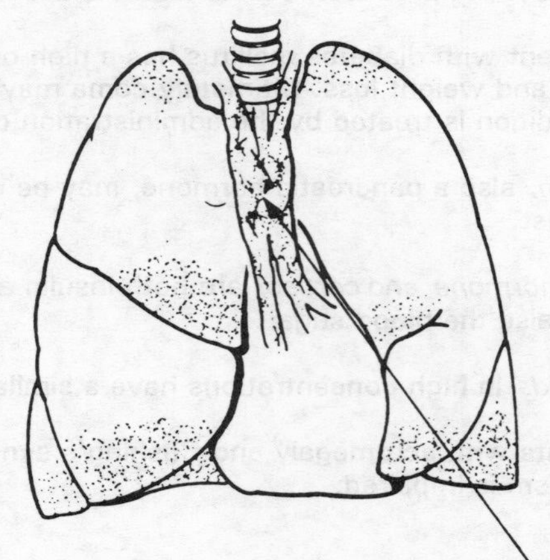

It becomes smaller after puberty and in adult life.

The thymus has a **cortex** which is packed full of lymphocytes, and a **medulla** composed of a loose mass of epithelial cells. The epithelial cells produce a factor — *the thymic humoric factor* — which stimulates lymphoid cells throughout the body to divide and develop the ability to recognise and attack foreign material.

Many responses to foreign material — notably the response to transplanted tissue and many infections — are mediated, not by soluble circulating antibodies but by cells. The cells involved are lymphocytes. The original development of these cells is within the thymus in embryonic life and early infancy. From here they migrate to become established in lymphoid tissue throughout the body. At this stage the thymus is essential to survival. Death from infection follows its removal. It continues to provide a minor source of lymphocytes and produce its factor after this time, but after childhood the lymphoid system is established and removal of the thymus results in little impairment of immunity.

TEST FOUR

1. **(a) Which element is incorporated into the thyroid hormones?**

   (b) Why is thyroid hormone stored?

   ____

   ____

2. **Which of the statements on the right apply to the items listed on the left?**

(i) Cretinism.	(a) A hormone produced within the thyroid which alters blood calcium levels.
(ii) Calcitonin.	(b) A condition caused by lack of thyroid hormone in an adult.
(iii) Myxoedema.	(c) A condition caused by lack of thyroid hormone in a child.

3. **If parathormone is involved in controlling of the body's calcium, why is excess secretion of the parathyroid glands characterised by decalcification of the bones?**

   ____

   ____

   ____

4. **Place ticks in the appropriate brackets to indicate the properties of the hormones.**

	Insulin	Glucagon	Growth hormone
(a) Increases blood sugar levels.	( )	( )	( )
(b) Decreases blood sugar levels.	( )	( )	( )
(c) Decreases insulin production.	( )	( )	( )

5. **In which of the following is the thymus important:**

   (a) in the production of antibodies?

   (b) in the development of lymphocytes?

   (c) in the production of lymphocytes in the adult?

ANSWERS TO TEST FOUR

1. (a) Iodine.

   (b) Because iodine may be absent from the diet for long periods.

2. (i) Cretinism — (c) is a condition caused by lack of thyroid hormones in a child.
   (ii) Calcitonin — (a) is a hormone produced within the thyroid which alters blood calcium levels.
   (iii) Myxoedema — (b) is a condition caused by lack of thyroid hormone in an adult.

3. Parathormone is involved in controlling the calcium ion concentration of the blood, so that excess secretion of the parathyroid glands greatly raises the concentration of blood calcium by depleting that of the bones.

4.

		Insulin	Glucagon	Growth hormone
(a)	Increases blood sugar levels.	( )	(√)	(√)
(b)	Decreases blood sugar levels.	(√)	( )	( )
(c)	Decreases insulin production.	( )	(√)	( )

5. (b) In the development of lymphocytes.

POST TEST

1. Label the endocrine glands indicated on the diagram alongside.

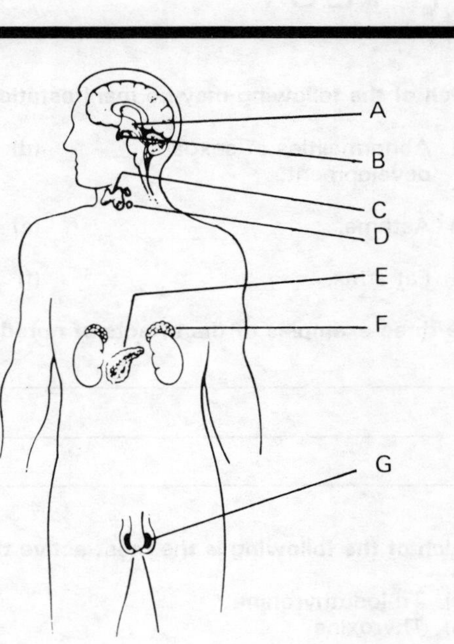

2. Complete the following:

(a) ________ is an alternative name for antidiuretic hormone.
(b) ADH is released into the blood by the________ gland together with a second hormone, ________.
(c) ADH helps maintain the ________of the blood at a ________ level.
(d) Diabetes insipidus refers to a condition caused by ________ of ADH in which volumes of ________ urine are constantly passed.

3. Which of the statements on the right apply to the hormones listed on the left?

(i) Aldosterone.	(a) Release is controlled by ACTH.
(ii) Cortisol.	(b) Release is controlled by nerve impulses.
(iii) Adrenalin.	(c) Release is controlled by renin release.

4. Which of the following are characteristics of Addison's disease?

(a) Low blood pressure.

(b) High blood pressure.

(c) Muscular weakness.

(d) Intolerance of stress.

(e) Increased cortisol production.

POST TEST

5. Which of the following may be manifestations of increased cortisol production?

(a) Abnormalities of sexual development.

(b) Asthma.

(c) Fat arms.

(d) Muscle wasting.

(e) Diabetes mellitus.

(f) Overproduction of red cells.

6. Give three examples of the effects of noradrenalin and adrenalin.

____

____

____

7. Which of the following is the most active thyroid hormone?

(a) Tri-iodothyronine.
(b) Thyroxine.
(c) Tyrosine.

8. Which of the symptoms on the right are features of the conditions listed on the left?

(i) Myxoedema.	(a) Subnormal temperature.
(ii) Hyperthyroidism	(b) Spontaneous fractures.
(iii) Parathyroid overproduction.	(c) Muscle spasm.
(iv) Parathyroid underproduction.	(d) Increased appetite.

9. Place ticks in the appropriate brackets to indicate which of the effects listed below are caused by thyroid overactivity and which are caused by parathyroid underactivity.

	Thyroid overactivity	Parathyroid underactivity
(a) Staring expression.	( )	( )
(b) Weight loss.	( )	( )
(c) Pins and needles.	( )	( )
(d) Tremor of the hands.	( )	( )
(e) Urinary stones.	( )	( )

10. Are the following statements true or false?

	True	False
(a) Insulin increases protein synthesis.	( )	( )
(b) Alpha cells secrete glucagon.	( )	( )
(c) Growth hormone increases the uptake of glucose by cells.	( )	( )
(d) Cushing's syndrome can cause diabetes mellitus.	( )	( )
(e) Glucagon causes glycogen mobilisation from the liver.	( )	( )

ANSWERS TO POST TEST

1. A Hypothalamus.
B Pituitary gland.
C Thyroid gland.
D Parathyroid gland.
E Pancreas.
F Suprarenal gland.
G Testis.

2. (a) *Vasopressin* is an alternative name for antidiuretic hormone.

(b) ADH is released into the blood by the *posterior pituitary* gland together with a second hormone, *oxytocin.*

(c) ADH helps maintain the *osmotic pressure* of the blood at a *constant* level.

(d) Diabetes insipidus refers to a condition caused by *deficiency* of ADH in which *large* volumes of *dilute* urine are constantly passed.

3. (i) Aldosterone (c) release is controlled by renin release.

(ii) Cortisol (a) release is controlled by ACTH.

(iii) Adrenalin (b) release is controlled by nerve impulses.

4. (a) Low blood pressure.

(c) Muscular weakness.

(d) Intolerance of stress.

ANSWERS TO POST TEST

5. (d) Muscle wasting.

(e) Diabetes mellitus.

(f) Overproduction of red cells.

6. Three of the following:

Diversion of blood to limb muscles.
Increase in the activity of the heart.
Dilatation of the bronchial tree.
Mobilisation of liver glycogen.
Reduction of activity in the gut.

7. (a) Tri-iodothyronine.

8.

(i)	Myxoedema	(a)	causes subnormal temperature.
(ii)	Hyperthyroidism	(d)	causes increased appetite.
(iii)	Parathyroid overproduction	(b)	causes spontaneous fractures.
(iv)	Parathyroid underproduction	(c)	causes muscle spasm.

9.

		Thyroid overactivity	Parathyroid underactivity
(a)	Staring expression.	(√)	( )
(b)	Weight loss.	(√)	( )
(c)	Pins and needles.	( )	(√)
(d)	Tremor of the hands.	(√)	( )
(e)	Urinary stones.	( )	(√)

10.

		True	False
(a)	Insulin increases protein synthesis.	(√)	( )
(b)	Alpha cells secrete glucagon.	(√)	( )
(c)	Growth hormone increases the uptake of glucose by cells.	( )	(√)
(d)	Cushing's syndrome can cause diabetes mellitus.	(√)	( )
(e)	Glucagon causes glycogen mobilisation from the liver.	(√)	( )

Contents

The Nervous System

1	INTRODUCTION	39
1.1	The basic plan of the nervous system	39
1.2	The neurone	40
1.3	The nerve impulse	42
1.4	The transmission of nerve impulses	44
1.5	The neuromuscular junction	45
	TEST ONE	47
	ANSWERS TO TEST ONE	48
2	THE SPINAL CORD AND THE SPINAL NERVES	49
2.1	Introduction	49
2.2	The structure of the spinal cord	50
2.3	The spinal nerves	52
2.4	Nervous reflexes	54
	TEST TWO	55
	ANSWERS TO TEST TWO	56
3	THE BRAIN	57
3.1	The development and divisions of the brain	57
3.2	The cerebral hemispheres	58
3.3	A vertical section through the brain	60
3.4	The brain stem	61
3.5	The reticular formation	62
3.6	The brain viewed from below	63
3.7	The cranial nerves	64
3.8	Motor pathways	66
3.9	Sensory pathways	67
3.10	The coverings of the brain and spinal cord	68
3.11	The blood supply of the brain	70
3.12	The functions of the brain	70
	TEST THREE	71
	ANSWERS TO TEST THREE	72
4	THE AUTONOMIC NERVOUS SYSTEM	73
4.1	Introduction	73
4.2	The parasympathetic nervous system	74
4.3	The sympathetic nervous system	75
	TEST FOUR	77
	ANSWERS TO TEST FOUR	78
	POST TEST	79
	ANSWERS TO POST TEST	81

1. Introduction

1.1. The basic plan of the nervous system

The nervous system is a collection of vast numbers of nerve cells, or *neurones.* These are cells with long branching processes (nerve fibres) which can conduct nerve impulses.

For convenience of description the nervous system is divided into two parts:

1. *the central nervous system (CNS)* which consists of:

   the two **cerebral hemispheres** (cerebrum),

   the **cerebellum,**
   with the **brain stem,**
   and the **spinal cord;**

2. *the peripheral nervous system* which consists of bundles of nerve fibres emerging from the CNS to pass to the rest of the body as peripheral nerves.

   There are:
   the **cranial nerves**
   12 pairs of nerves arising from the brain stem to supply mainly the head and neck;

   the **spinal nerves**
   31 pairs of nerves arising from the spinal cord to supply the trunk and limbs.

   The spinal nerves form **limb plexuses** where bundles of nerves intermingle.

Functionally, the nervous system is divided into:

the somatic (or voluntary) nervous system which is associated with impulses to the limbs and body wall; and

the autonomic (or involuntary) *nervous system* which is associated with impulses to the viscera and blood vessels.

Many individual neurones cross the boundaries of these divisions and the nervous system functions in a co-ordinated and unified manner.

1.2. The neurone

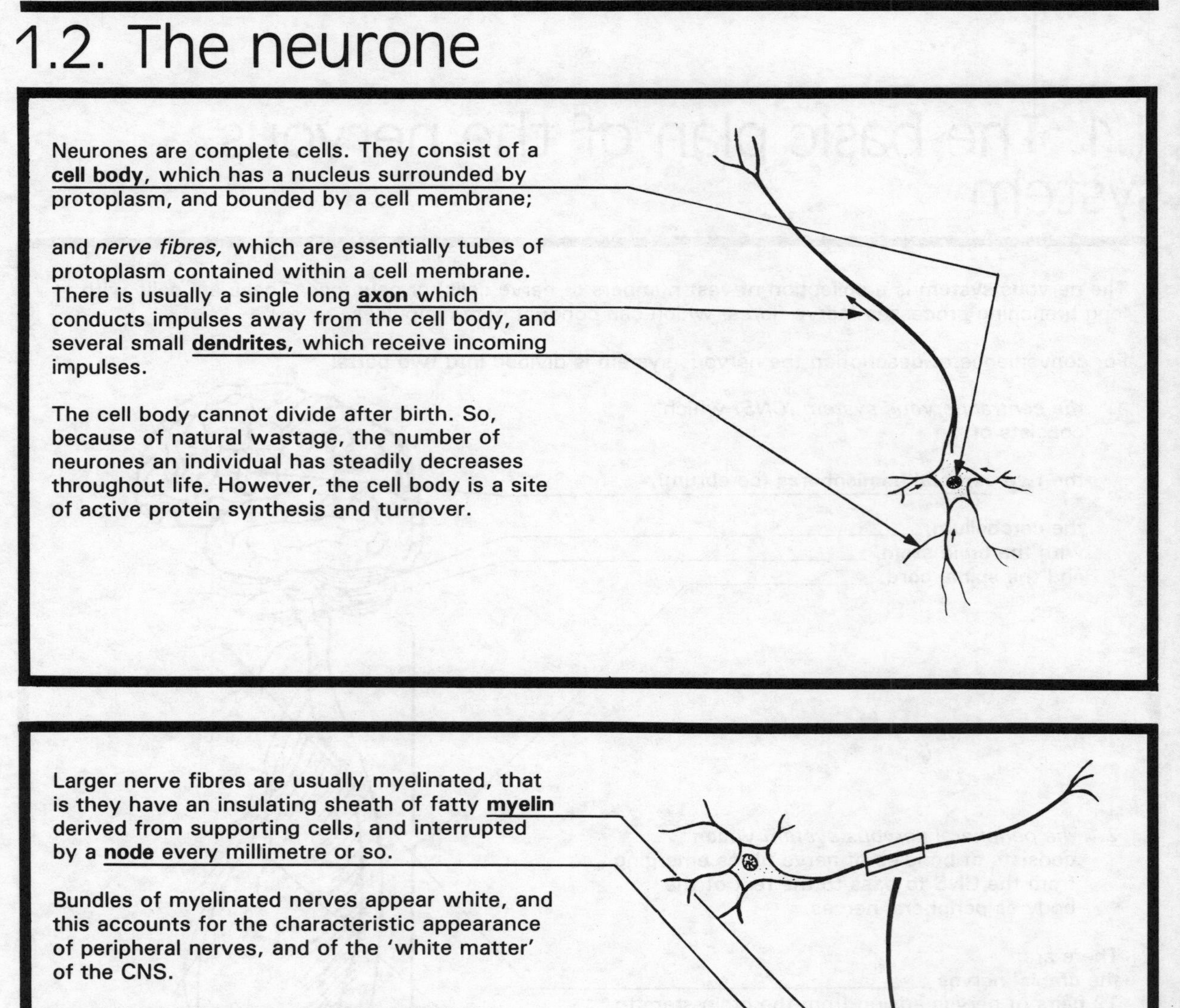

Neurones are complete cells. They consist of a **cell body**, which has a nucleus surrounded by protoplasm, and bounded by a cell membrane;

and *nerve fibres,* which are essentially tubes of protoplasm contained within a cell membrane. There is usually a single long **axon** which conducts impulses away from the cell body, and several small **dendrites,** which receive incoming impulses.

The cell body cannot divide after birth. So, because of natural wastage, the number of neurones an individual has steadily decreases throughout life. However, the cell body is a site of active protein synthesis and turnover.

Larger nerve fibres are usually myelinated, that is they have an insulating sheath of fatty **myelin** derived from supporting cells, and interrupted by a **node** every millimetre or so.

Bundles of myelinated nerves appear white, and this accounts for the characteristic appearance of peripheral nerves, and of the 'white matter' of the CNS.

Smaller nerves (with a diameter less than 2 μmm) may not possess a myelin sheath and they then appear dull grey.

Peripheral nerves are surrounded by specialised *Schwann cells.* These produce the myelin sheath of myelinated nerves. They also permit repair of nerve fibres. When a nerve is cut the axon and sheath beyond the cut disintegrate. The Schwann cells, however, persist as a tunnel so that as the central stump sprouts and grows, the new fibre can find its way back to the original muscle, gland or sense organ, over long distances.

There are no Schwann cells in the CNS. No repair is therefore possible, and damage within the CNS is permanent.

Within the CNS the neurones are supported and nourished by specialised *glial cells.* There is no connective tissue at all, so CNS tissue is very soft.

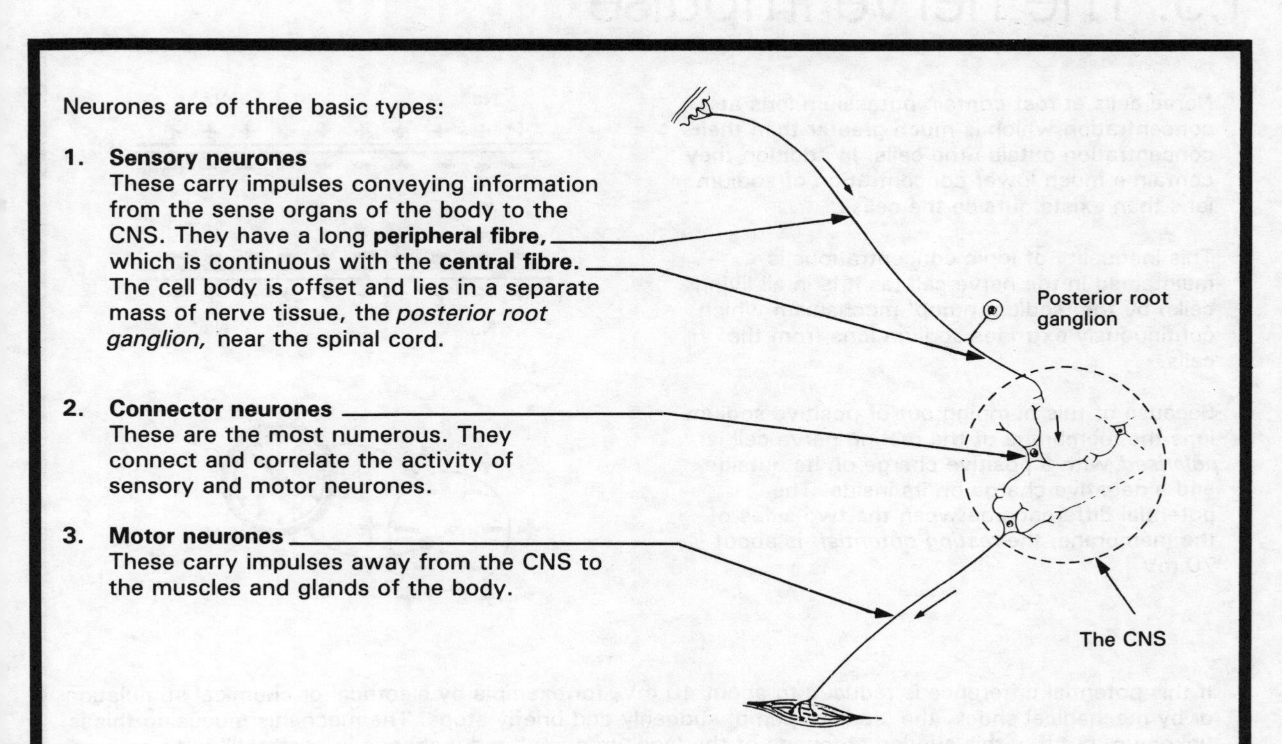

Neurones are of three basic types:

1. **Sensory neurones**
   These carry impulses conveying information from the sense organs of the body to the CNS. They have a long **peripheral fibre**, which is continuous with the **central fibre**. The cell body is offset and lies in a separate mass of nerve tissue, the *posterior root ganglion,* near the spinal cord.

2. **Connector neurones**
   These are the most numerous. They connect and correlate the activity of sensory and motor neurones.

3. **Motor neurones**
   These carry impulses away from the CNS to the muscles and glands of the body.

Co-ordination between sensory and motor nerves may be very simple, occurring within the spinal cord.

More complex responses to incoming sensation are possible by the connector neurone actively involving the 'higher centres' of the brain stem and cerebellum.

The most complex reactions involve the *cortex of the cerebral hemispheres.* 90% of all neurones are here and they co-ordinate responses involving such factors as reasoning, memory and emotion.

1.3. The nerve impulse

Nerve cells at rest contain potassium ions at a concentration which is much greater than their concentration outside the cells. In addition they contain a much lower concentration of sodium ions than exists outside the cells.

This inequality of ionic concentrations is maintained in the nerve cell (as it is in all living cells) by the 'sodium pump' mechanism which continuously extrudes sodium ions from the cells.

Because of this pumping out of positive sodium ions the membrane of the resting nerve cell is *polarised* with a positive charge on its outside and a negative charge on its inside. The potential difference between the two sides of the membrane, the *resting potential,* is about 70 mV.

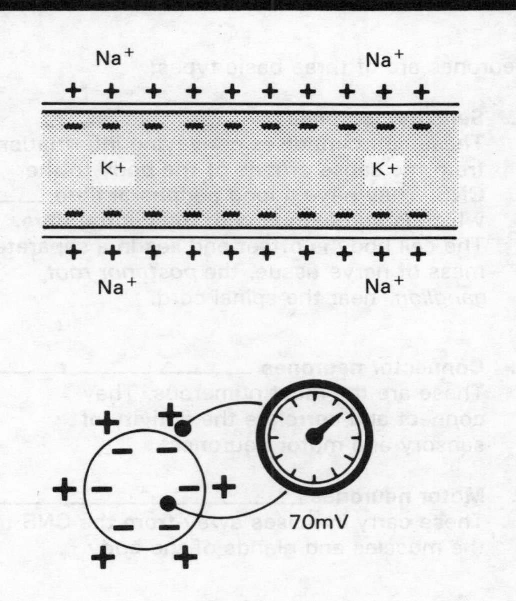

If this potential difference is reduced to about 40 mV, for example by electrical or chemical stimulation or by mechanical shock, the 'sodium pump' suddenly and briefly stops. The mechanism causing this is unknown, but it is this sudden stoppage of the 'sodium pump' and a change in permeability to sodium ions of the cell membrane which starts the propagation of the nerve impulse.

Sodium ions flood into the cell and cause the polarity of part of the membrane to be reversed.

Electric currents flow between the parts of the membrane which are at different potentials.

This electrical activity excites the adjacent part of the nerve fibre in turn and so a wave of polarity reversal is propagated along the fibre. This is the nerve impulse.

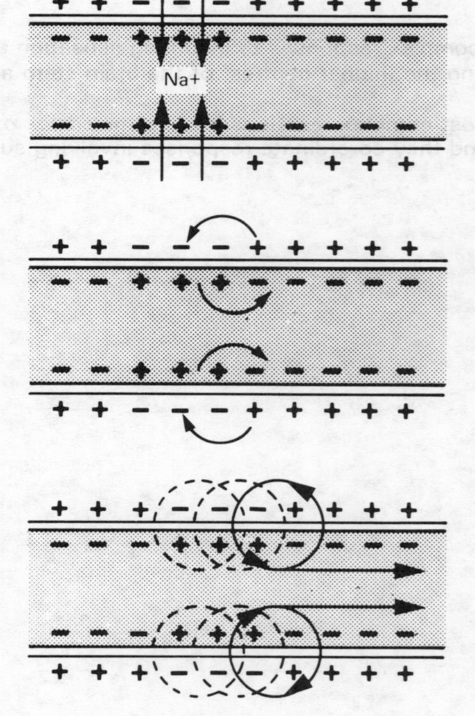

Almost immediately after the 'sodium pump' has stopped working, potassium leaks outward across the cell membrane. This rapidly restores the resting potential. Soon afterwards the 'sodium pump' resumes operation, and the usual distribution of electrolytes is restored.

The changes which occur when the nerve cell membrane is stimulated can be measured. The graph of electrical potential against time which displays the changes is known as the *action potential.*

The main features of the action potential curve are:

(i) a gradual rise at first due to the initial stimulus;

(ii) a rapid rise due to the inward rush of sodium ions;

(iii) a fall due to the outward flow of potassium ions;

(iv) the restoration of the resting potential.

+20
+10
0
−10
−20
−30
−40
−50
−60
1
2
3
4
Time (millisecs)

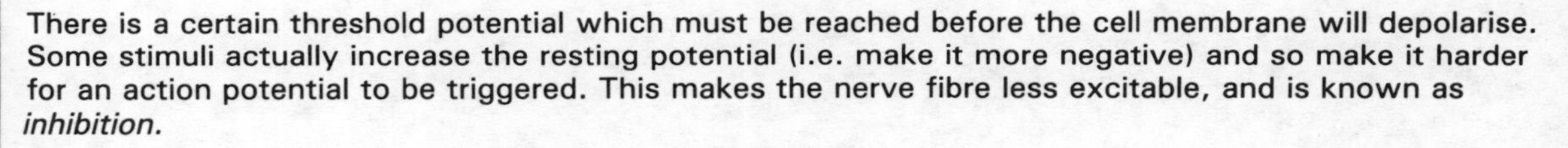

There is a certain threshold potential which must be reached before the cell membrane will depolarise. Some stimuli actually increase the resting potential (i.e. make it more negative) and so make it harder for an action potential to be triggered. This makes the nerve fibre less excitable, and is known as *inhibition.*

The speed of conduction of the nerve impulse depends on:

(a) the diameter of the nerve fibre; the greater the diameter the greater the speed;

(b) whether or not the fibre is myelinated; conduction is faster in myelinated fibres.

In myelinated fibres the local current flow is not to the adjacent part of the cell membrane, but electrostatically over the outside of the myelin sheath from one node to the next.

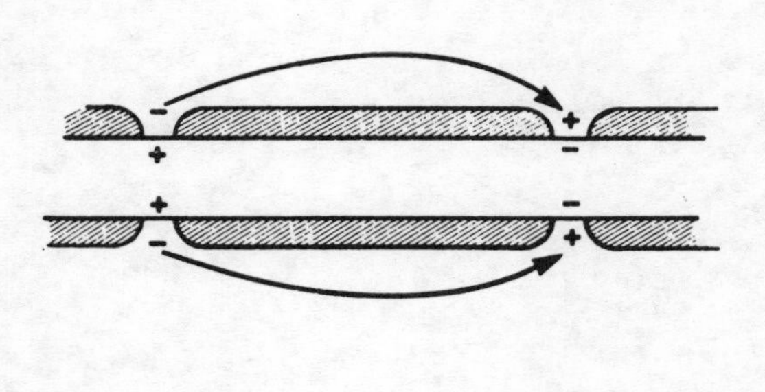

1.4. The transmission of nerve impulses

Nerve impulses are transmitted from one neurone to another at junctions known as *synapses.*

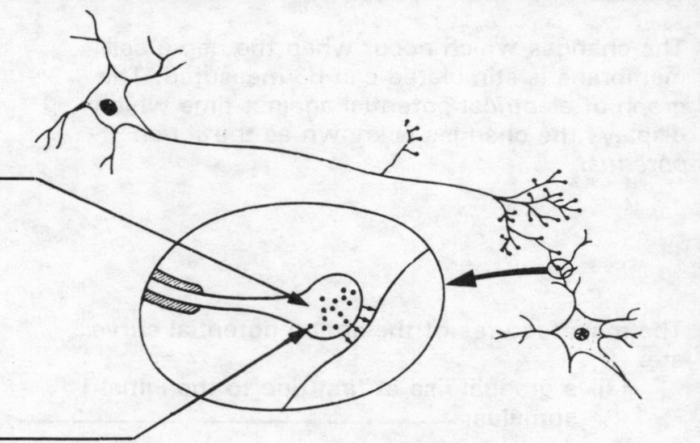

At a synapse the **boutons terminaux,** at the end of the branches of the axon of one nerve cell, lie against the dendrites or cell body of another nerve cell. There is no direct protoplasmic union across the synapse.

When a nerve impulse arrives at the end of the axon it causes molecules of a chemical transmitter substance to be released from **vesicles** in the boutons.

The chemical transmitter diffuses across the synaptic cleft and makes the membrane of the adjacent cell more permeable to sodium ions. If sufficient boutons are stimulated enough chemical transmitter may be released to depress the polarity of the whole cell body of the adjacent cell and so an action potential may be generated there.

Acetylcholine is the chemical transmitter in many synapses. A number of other substances are also known to act as chemical transmitters within the central nervous system. These include noradrenalin, dopamine, histamine and 5-Hydroxytryptamine and GABA.

Some boutons release chemical transmitters which make the adjacent nerve cell membrane *less* permeable to sodium ions. In so doing they inhibit the transmission of the nerve impulse. γ-aminobutyric acid (GABA) is probably such a transmitter. Certain drugs have been derived from GABA which inhibit the nervous system and help to overcome muscle spasm due to disease affecting the nervous system.

1.5. The neuromuscular junction

A typical motor neurone has a large cell body which lies inside the central nervous system. Many thousands of boutons terminate close to the surface of the cell body from many different sources.

Some of the boutons produce excitatory effects, some inhibitory effects. The balance of their action determines whether the nerve transmits an impulse (fires) or not.

The **axon** of the motor neurone leaves the central nervous system and runs in a peripheral nerve to its muscle where it branches to supply a scattered group of muscle fibres within the muscle — its *motor unit.*

Some motor units consist of only 5–10 fibres. These are able to provide very fine movements, as in the muscles of the eye. Coarser muscles, for example those of the back, may contain motor units consisting of thousands of fibres.

The nerve fibre terminates near the centre of each muscle fibre as a **motor end plate.** This is the neuromuscular junction.

The motor end plate contains many vesicles of acetylcholine. An incoming nerve impulse causes some acetylcholine to be released. The acetylcholine diffuses across the synaptic cleft and acts on the surface of the muscle cell membrane to depolarise it. An action potential is generated in the muscle fibre and travels along the fibre, stimulating it to contract.

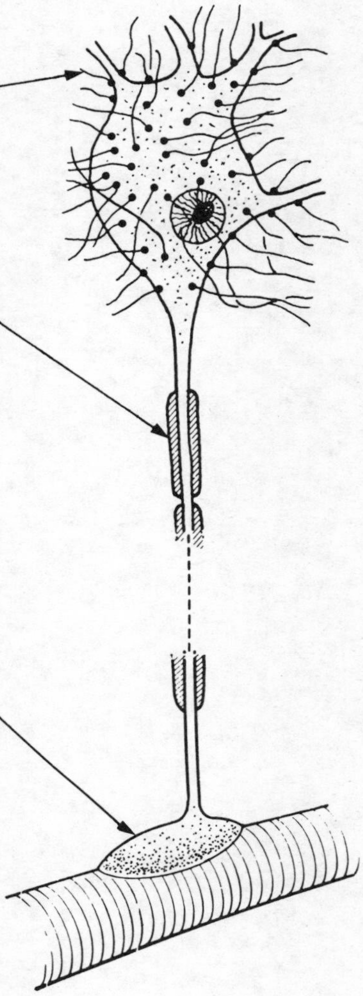

The motor neurone and its own muscle fibres form an intimate functional unit. If the cell body dies or the fibre is cut and fails to re-grow back to the muscle, then the muscle fibres rapidly waste away. If this should happen to a growing limb on a large scale (for example in infantile poliomyelitis, in which a virus selectively injures motor neurones), then growth of the entire limb ceases. This does not happen if the motor neurone is cut off from 'higher centres' (for example in spinal cord injury).

TEST ONE

1. Indicate which of the following structures form part of the central nervous system, and which belong to the peripheral nervous system by placing ticks in the appropriate brackets.

		Central nervous system	Peripheral nervous system
(a)	Spinal cord.	( )	( )
(b)	Cranial nerves.	( )	( )
(c)	Brain.	( )	( )
(d)	Spinal nerves.	( )	( )
(e)	Nerve plexuses.	( )	( )

2. Which of the descriptions on the right apply to the types of neurone listed on the left?

(i) A sensory neurone.
(ii) A motor neurone.
(iii) A connector neurone.

(a) Connects the other two neurones.
(b) Carries impulses away from the CNS.
(c) Carries impulses towards the CNS.
(d) Has its cell body outside the CNS.
(e) Carries impulses towards muscles or glands.
(f) Is the most common type of neurone.

3. What is myelin, and where is it found?

____

____

4. Place the following events in the order in which they occur.

(a) The nerve fibre is polarised with a positive charge on its inside.

(b) Sodium ions flood into the nerve fibre.

(c) The sodium pump briefly ceases operating.

(d) Potassium leaks out of the nerve fibre.

(e) The nerve fibre is polarised with a negative charge on its inside.

(f) The nerve fibre is partly depolarised by some means.

5. (a) What kind of substance is acetylcholine?

____

(b) Name other similar substances.

____

6. Which of the following describes a motor unit?

(a) A group of motor neurones.

(b) A group of muscles.

(c) A group of muscle fibres.

ANSWERS TO TEST ONE

1.

		Central nervous system	Peripheral nervous system
(a)	Spinal cord.	(√)	( )
(b)	Cranial nerves.	( )	(√)
(c)	Brain.	(√)	( )
(d)	Spinal nerves.	( )	(√)
(e)	Nerve plexuses.	( )	(√)

2. (i) A sensory neurone (c) carries impulses towards the CNS, and (d) has its cell body outside the CNS.

(ii) A motor neurone (b) carries impulses away from the CNS, and (e) carries impulses towards muscles or glands.

(iii) A connector neurone (a) connects the other two neurones, and (f) is the most common type of neurone.

5. Myelin is a white fatty substance found insulating the larger nerve fibres of the CNS.

4. 1. (e) The nerve fibre is polarised with a negative charge on its inside.

2. (f) The nerve fibre is partly depolarised by some means.

3. (c) The sodium pump briefly ceases operating.

4. (b) Sodium ions flood into the nerve fibre.

5. (a) The nerve fibre is polarised with a positive charge on its inside.

6. (d) Potassium leaks out of the nerve fibre.

5. (a) Acetylcholine is a chemical transmitter within the central nervous system.

(b) Other such transmitters include noradrenalin, dopamine, histamine, 5-hydroxytryptamine and GABA.

6. (a) A group of motor neurones.

(c) A group of muscle fibres.

2. The spinal cord and the spinal nerves

2.1. Introduction

The **spinal cord** floats in cerebrospinal fluid within the spinal canal of the vertebral column. It extends from the foramen magnum of the skull to the second lumbar intervertebral disc.

The **spinal nerves** leave the spinal canal above and below the pedicles of adjacent vertebrae.

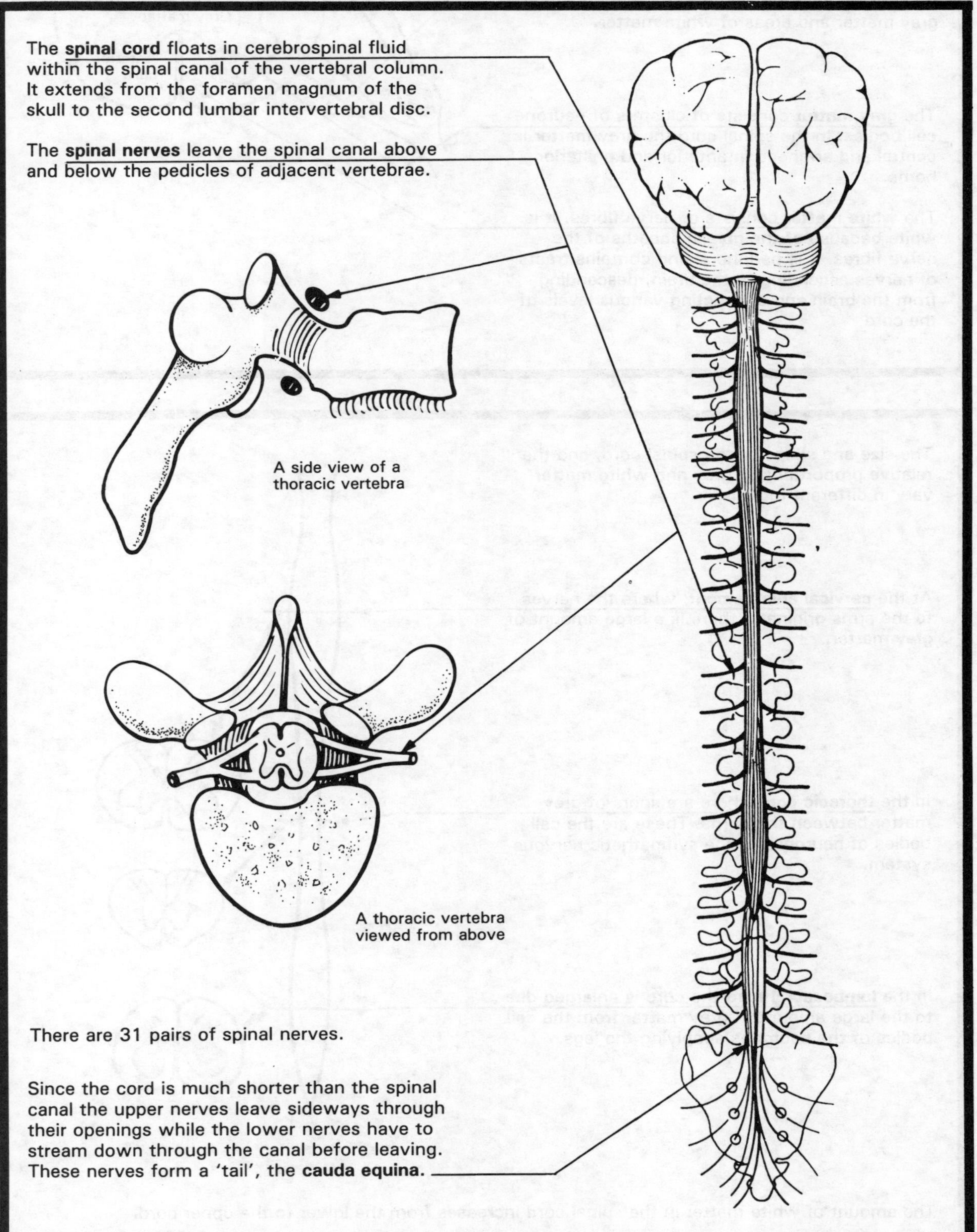

A side view of a thoracic vertebra

A thoracic vertebra viewed from above

There are 31 pairs of spinal nerves.

Since the cord is much shorter than the spinal canal the upper nerves leave sideways through their openings while the lower nerves have to stream down through the canal before leaving. These nerves form a 'tail', the **cauda equina**.

2.2. The structure of the spinal cord

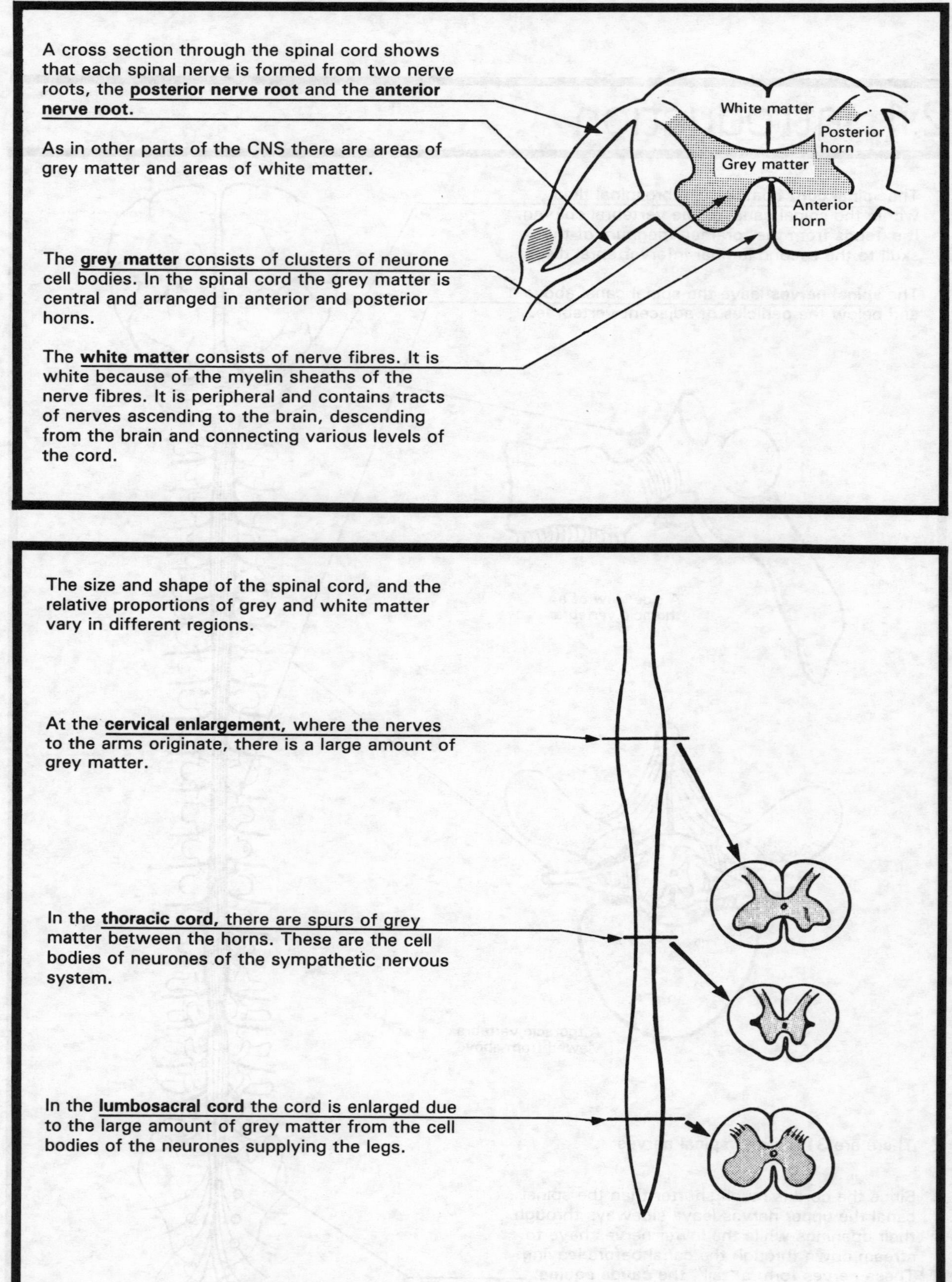

A cross section through the spinal cord shows that each spinal nerve is formed from two nerve roots, the **posterior nerve root** and the **anterior nerve root.**

As in other parts of the CNS there are areas of grey matter and areas of white matter.

The **grey matter** consists of clusters of neurone cell bodies. In the spinal cord the grey matter is central and arranged in anterior and posterior horns.

The **white matter** consists of nerve fibres. It is white because of the myelin sheaths of the nerve fibres. It is peripheral and contains tracts of nerves ascending to the brain, descending from the brain and connecting various levels of the cord.

The size and shape of the spinal cord, and the relative proportions of grey and white matter vary in different regions.

At the **cervical enlargement,** where the nerves to the arms originate, there is a large amount of grey matter.

In the **thoracic cord,** there are spurs of grey matter between the horns. These are the cell bodies of neurones of the sympathetic nervous system.

In the **lumbosacral cord** the cord is enlarged due to the large amount of grey matter from the cell bodies of the neurones supplying the legs.

The amount of white matter in the spinal cord increases from the lower to the upper cord.

Within the white matter, nerve fibres travelling to similar destinations are grouped together to form *tracts*. These tracts are not visible but their presence is dramatically demonstrated by the effects of a local injury. Similarly, areas of specialisation exist in the grey matter.

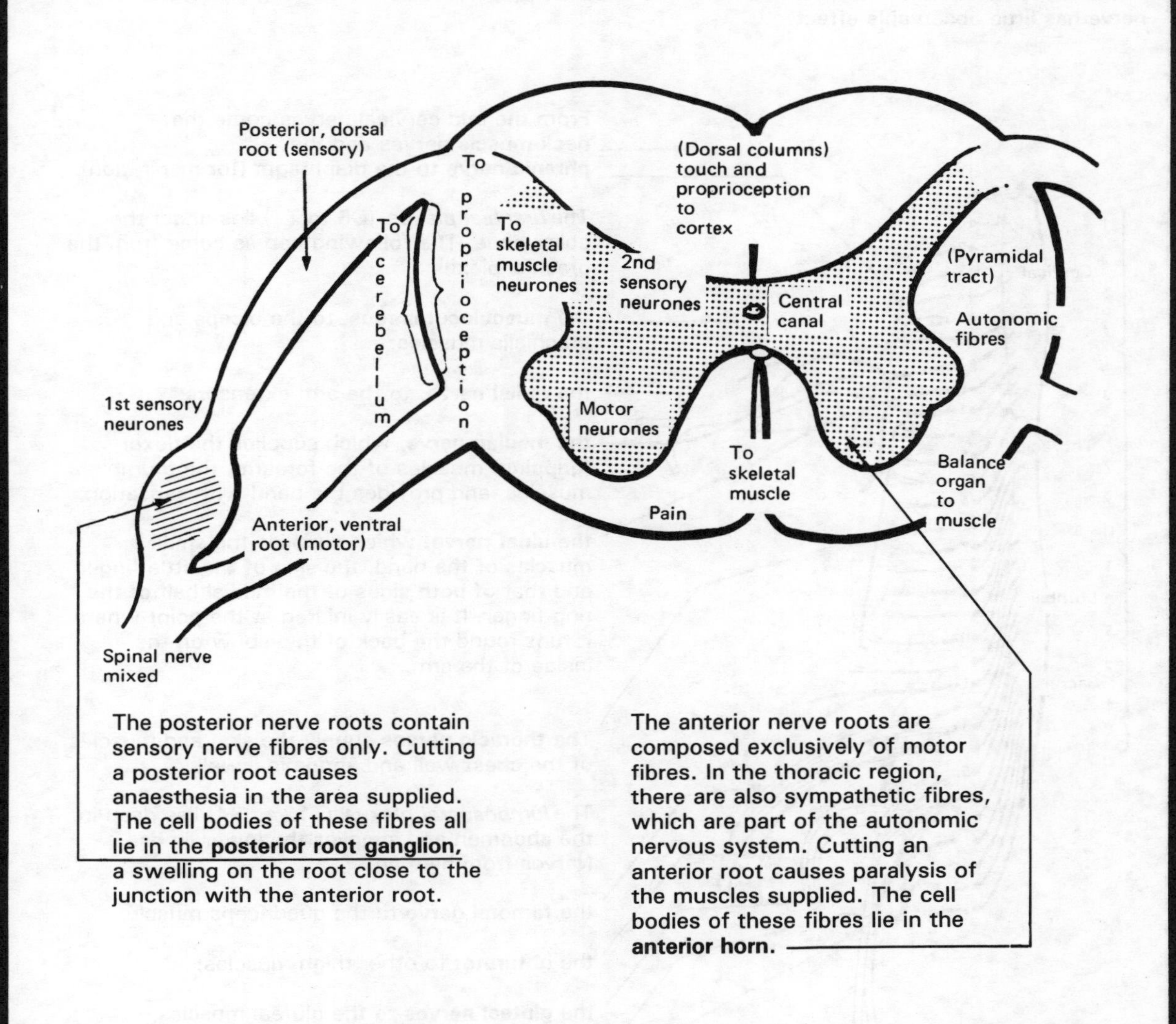

The posterior nerve roots contain sensory nerve fibres only. Cutting a posterior root causes anaesthesia in the area supplied. The cell bodies of these fibres all lie in the **posterior root ganglion,** a swelling on the root close to the junction with the anterior root.

The anterior nerve roots are composed exclusively of motor fibres. In the thoracic region, there are also sympathetic fibres, which are part of the autonomic nervous system. Cutting an anterior root causes paralysis of the muscles supplied. The cell bodies of these fibres lie in the **anterior horn.**

The spinal nerve formed by the fusion of the posterior and anterior nerve roots is a mixed nerve, containing both sensory and motor nerve fibres.

2.3. The spinal nerves

The spinal nerves are numbered according to the vertebrae at which they leave the spinal cord. Each nerve supplies a definite mass of muscles and a definite area of skin, for example the first thoracic nerve (T1) supplies the small muscles of the hand and an area of skin on the inner aspect of the arm. There is considerable overlap between the area supplied by nerves, so that cutting a single spinal nerve has little observable effect.

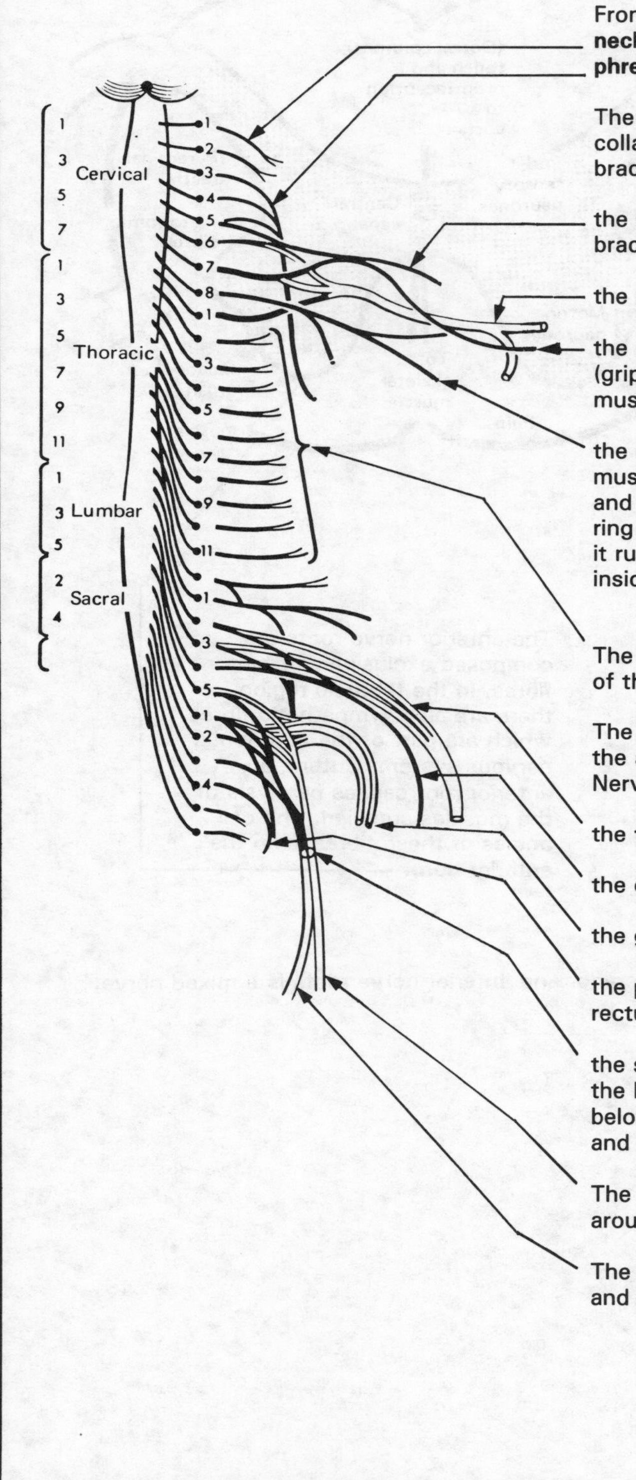

From the mid cervical nerves come the **neck muscle nerves** and the **phrenic nerve** to the diaphragm (for respiration).

The *brachial plexus* (C5 to T1) lies under the collar bone. The following nerves come from the brachial plexus:

the **musculocutaneous**, to the biceps and brachialis muscles;

the **radial nerve**, to the arm extensors;

the **median nerve**, which supplies the flexor (gripping) muscles of the forearm, the thumb muscles, and provides the hand with sensation;

the **ulnar nerve**, which supplies the small muscles of the hand, the skin of the little finger and that of both sides of the medial half of the ring finger. It is easily injured at the point where it runs round the back of the elbow on the inside of the arm.

The **thoracic nerves** supply the skin and muscles of the chest wall and abdominal wall.

The *lumbosacral plexus* (T12 to S4) lies deep in the abdomen and supplies the lower limbs. Nerves from here are:

the **femoral nerve** to the quadriceps muscle;

the **obturator** to other thigh muscles;

the **gluteal nerves** to the gluteal muscles;

the **pudendal nerve** which supplies the bladder, rectum and perineum.

the **sciatic nerve** which is the largest nerve in the body, supplying all the muscles and skin below the knee. It starts deep in the buttock and passes behind the knee into the leg.

The **common peroneal** branch is wrapped around the fibula. It supplies the foot extensors.

The **tibial branch** supplies the large calf muscles and sensation to most of the foot.

A peripheral nerve, such as the radial nerve, is made up of **nerve bundles,** sheathed in a **perineurium** of tough fibrous tissue.

Each bundle may contain hundreds of nerve fibres, sensory and motor, myelinated and unmyelinated, all mixed together.

The bundles give the nerve a faintly striped appearance. They branch when the nerve branches.

The bundles are enclosed in a thicker tough fibrous **epineurium.** This covering makes the peripheral nerves tough and resistant to injury by stretching, but the delicate protoplasmic axons are sensitive to prolonged pressure.

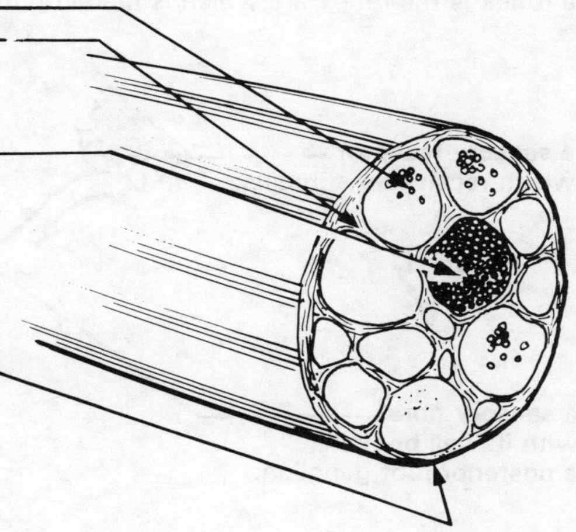

2.4. Nervous reflexes

The nervous reflex action is a rapid involuntary response to a sensory stimulus. The anatomical basis of a reflex is the reflex arc which is made up of the following components:

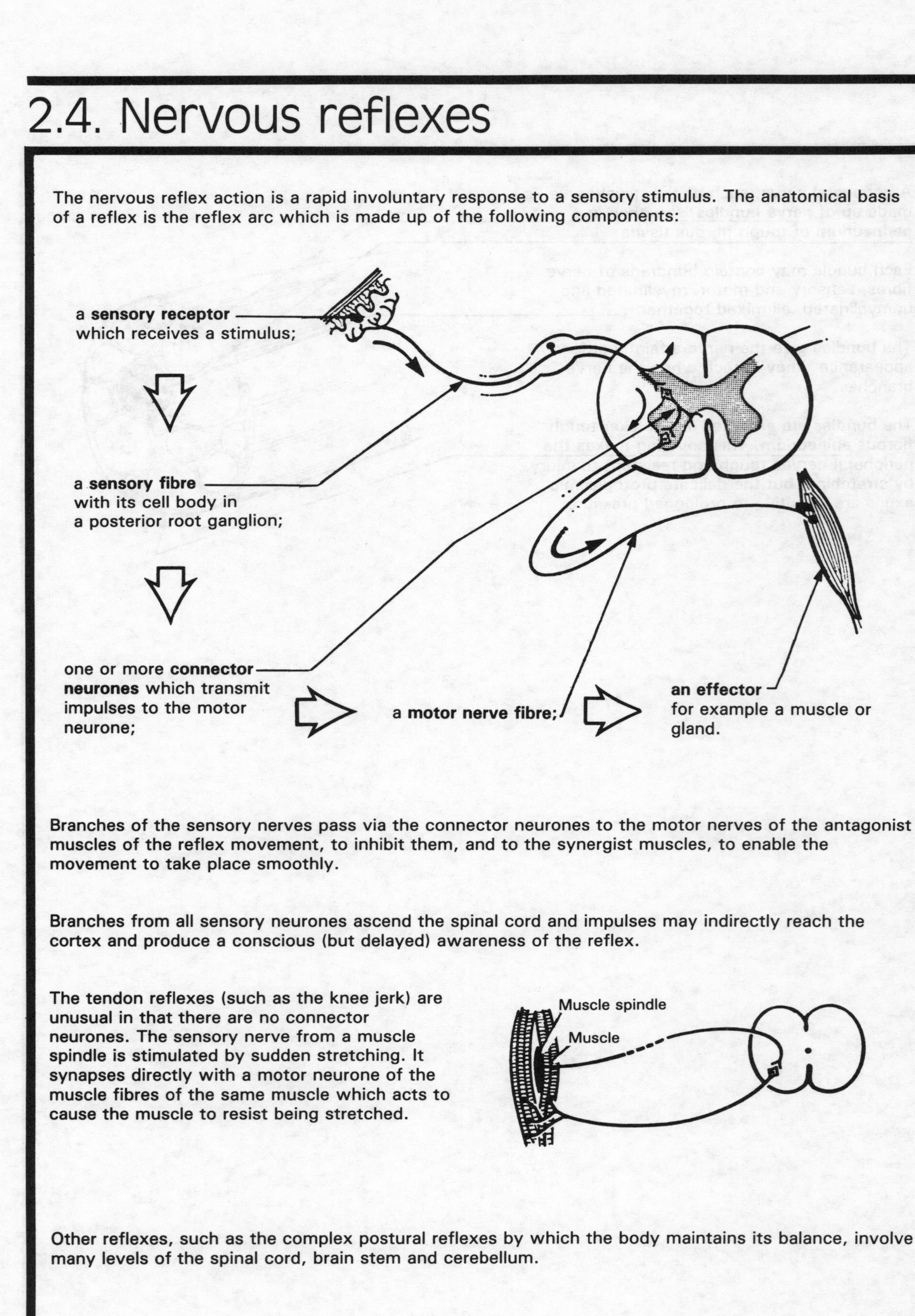

Branches of the sensory nerves pass via the connector neurones to the motor nerves of the antagonist muscles of the reflex movement, to inhibit them, and to the synergist muscles, to enable the movement to take place smoothly.

Branches from all sensory neurones ascend the spinal cord and impulses may indirectly reach the cortex and produce a conscious (but delayed) awareness of the reflex.

The tendon reflexes (such as the knee jerk) are unusual in that there are no connector neurones. The sensory nerve from a muscle spindle is stimulated by sudden stretching. It synapses directly with a motor neurone of the muscle fibres of the same muscle which acts to cause the muscle to resist being stretched.

Other reflexes, such as the complex postural reflexes by which the body maintains its balance, involve many levels of the spinal cord, brain stem and cerebellum.

TEST TWO

1. Name the vertebral level at which the spinal cord ends.

____

2. Indicate which of the names in the list below refer to the parts of the spinal cord labelled on the diagram alongside, by placing the appropriate letters in the brackets.

1. White matter. ( )
2. Posterior root. ( )
3. Grey matter. ( )
4. Anterior root. ( )
5. Posterior root ganglion. ( )

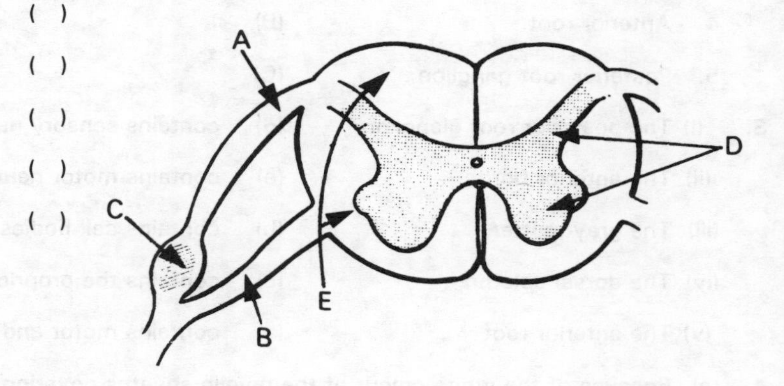

3. Which of the statements on the right apply to the items listed on the left?

(i) The posterior root ganglion.	(a) Contains motor and autonomic fibres.
(ii) The anterior horn.	(b) Contains cell bodies of all types.
(iii) The grey matter.	(c) Contains the proprioception fibres.
(iv) The dorsal column.	(d) Contains sensory neurone cell bodies.
(v) The anterior root.	(e) Contains motor neurone cell bodies.

4. (a) Why is white matter white?

____

(b) What is the effect of cutting the anterior nerve roots?

____

(c) Why is the anterior horn increased in bulk in the lumbar region?

____

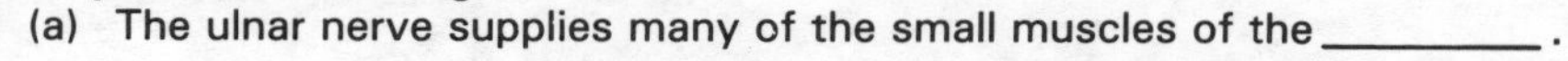

5. Complete the following:

(a) The ulnar nerve supplies many of the small muscles of the ________.

(b) The median nerve supplies those muscles of the ________ that are used in ________.

(c) The phrenic nerve supplies the ________ which is raised and lowered in ________.

(d) The sciatic nerve supplies ________ and muscle below the ________.

(e) The thoracic nerves supply ________ and muscles of the ________ and ________.

ANSWERS TO TEST TWO

1. The spinal cord ends at the second lumbar intervertebral disc.

2.

1.	White matter.	(E)
2.	Posterior root.	(A)
3.	Grey matter.	(D)
4.	Anterior root.	(B)
5.	Posterior root ganglion.	(C)

3.

(i) The posterior root glanglion	(d)	contains sensory neurone cell bodies.
(ii) The anterior horn	(e)	contains motor neurone cell bodies.
(iii) The grey matter	(b)	contains cell bodies of all types.
(iv) The dorsal column	(c)	contains the proprioception and touch fibres.
(v) The anterior root	(a)	contains motor and autonomic fibres

4. (a) Because of the white colour of the myelin sheaths covering the nerve fibres.

(b) Paralysis and wasting of muscles.

(c) Because of the concentration of cell bodies of nerve fibres supplying the lower limbs.

5. (a) The ulnar nerve supplies many of the small muscles of the *hand*.
(b) The median nerve supplies those muscles of the *forearm* and hand that are used in *gripping.*
(c) The phrenic nerve supplies the *diaphragm* which is raised and lowered in *breathing.*
(d) The sciatic nerve supplies *skin* and muscle below the *knee.*
(e) The thoracic nerves supply *skin* and muscles of the *chest* and *abdominal walls.*

3. The brain

3.1. The development and divisions of the brain

The brain and spinal cord develop from a neural plate which runs as a strip down the back of the embryo. This folds over to form a neural tube with a central cavity.

The cavity later forms the ventricles of the brain and the central canal of the spinal cord.

The enlarged end of the tube has *three* major divisions.

From the forebrain two hollow buds grow out. They enlarge enormously and become C-shaped to form the cerebral hemispheres. Eventually these completely envelope the rest of the forebrain. The cavities of the cerebral hemispheres are the lateral ventricles. The cavity of the forebrain is the third ventricle. The 'stalks' of the cerebral hemispheres become very thick forming the cerebral peduncles. Each cerebral hemisphere becomes a hollow mass of white matter surrounded by a thick coating of grey matter — *the cortex.*

The **cerebellum** begins as a strip of tissue on the surface of the brain, and eventually comes to cover an expansion of the central cavity, the **fourth ventricle.**

A band joining the two sides of the cerebellum, **the pons,** develops anteriorly.

The remainder of the hind brain is the **medulla.**

The two hemispheres arising from part of the forebrain, at first have no direct link with each other. Then a thickening appears in the forebrain between them, which develops into a thick band. This grows rapidly backwards, eventually arching under the forebrain and over the lateral ventricles as a thick connecting tract between the **hemispheres** — the **corpus callosum.**

The wall of the forebrain behind the 'stalks' becomes a mass of grey matter, **the thalamus.** The floor of the forebrain becomes the **hypothalamus.**

The inner wall of the primitive cerebral hemisphere adjacent to the cerebral peduncle and thalamus becomes a complex mass of grey matter — the **basal ganglia.**

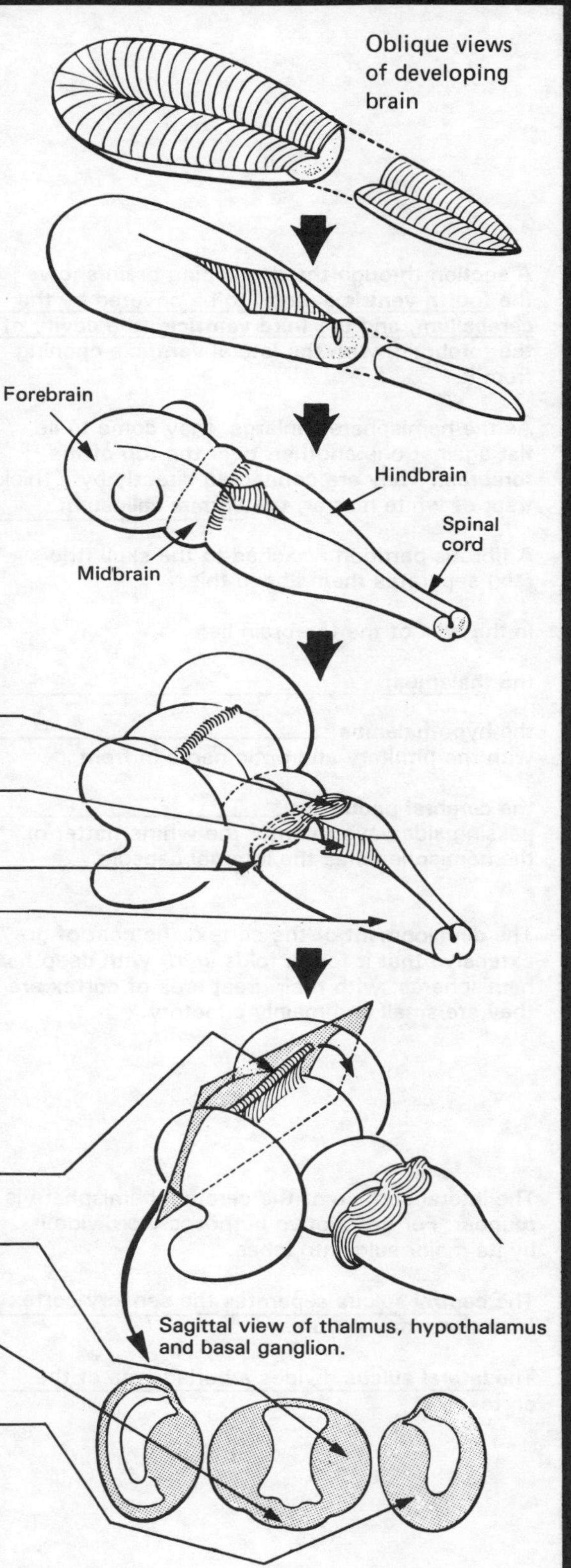

3.2. The cerebral hemispheres

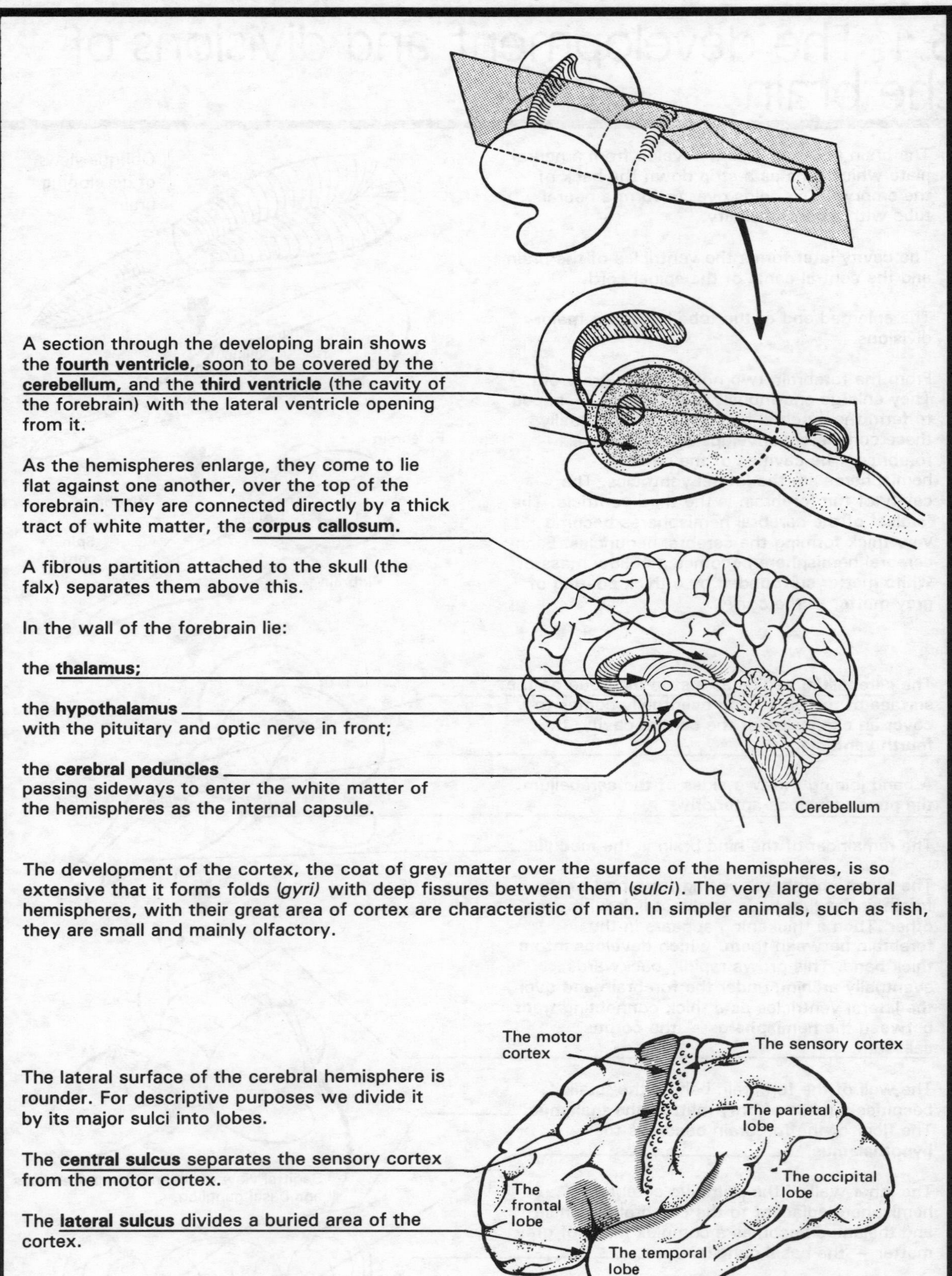

A section through the developing brain shows the **fourth ventricle,** soon to be covered by the **cerebellum,** and the **third ventricle** (the cavity of the forebrain) with the lateral ventricle opening from it.

As the hemispheres enlarge, they come to lie flat against one another, over the top of the forebrain. They are connected directly by a thick tract of white matter, the **corpus callosum.**

A fibrous partition attached to the skull (the falx) separates them above this.

In the wall of the forebrain lie:

the **thalamus;**

the **hypothalamus** with the pituitary and optic nerve in front;

the **cerebral peduncles** passing sideways to enter the white matter of the hemispheres as the internal capsule.

The development of the cortex, the coat of grey matter over the surface of the hemispheres, is so extensive that it forms folds (*gyri*) with deep fissures between them (*sulci*). The very large cerebral hemispheres, with their great area of cortex are characteristic of man. In simpler animals, such as fish, they are small and mainly olfactory.

The lateral surface of the cerebral hemisphere is rounder. For descriptive purposes we divide it by its major sulci into lobes.

The **central sulcus** separates the sensory cortex from the motor cortex.

The **lateral sulcus** divides a buried area of the cortex.

Certain areas of the cerebral hemispheres are specifically associated with certain functions:

The **motor cortex**, like the sensory cortex, is divided into areas representing the parts of the body whose movement it controls. The hand, especially the thumb, and the facial muscles have a larger representation than other muscle groups. Stimulation *always* causes complete movements, with prime movers, synergists and antagonists all working. Damage causes paralysis of willed movement, although reflex and involuntary movement can still occur.

The **sensory cortex** is divided up into areas representing areas of the body surface (on the opposite side). The important sense appreciating areas, such as the hand and the face, have a much larger representation than less sensitive areas such as the back. Damage to the sensory cortex makes the subject unable to identify the exact site of stimulation although still aware of the stimulus.

The **auditory cortex** receives impulses from both ears not just the opposite side as with most sensations.

The **parietal cortex** is concerned with the higher appreciation of sensation, such as the ability to assess the weight, texture and identity of an object.

The **visual cortex** is mainly on the inner side of the occipital lobe. It receives light impulses from the opposite visual field. Damage causes blindness.

The **speech cortex**, where thoughts are formulated into words, lies in the left hemisphere in most subjects (i.e. those that are right handed). This hemisphere is known as the dominant hemisphere as it also contains the motor cortex of the right hand. The non-dominant hemisphere probably contains equivalent areas dealing with visual rather than verbal information.

Damage to the dominant hemisphere, or to its connections to the brain stem causes the subject to lose the ability to speak sensibly, and may cause confusion since verbalisation is so important in enabling one to order one's thoughts.

Speech, heard by the auditory cortex is decoded and understood in this area.

The **temporal lobe** is largely a sensory association area. Impulses concerned with what is seen, heard, felt, etc., are co-ordinated and understood here. Local injury can cause strange behaviour associated with complex visual and auditory hallucinations.

3.3. A vertical section through the brain

A vertical section through the brain shows the following features.

Lateral ventricle

The cortex has a complex layered structure of cell bodies and fibres. It is about 5 mm thick.

The white matter of the cortex is a mass of interlaced fibres connecting the cortical neurones with each other and with the brain stem in an incredibly complex pattern.

The **corpus callosum** is a mass of white matter which enables the hemispheres to correlate with one another.

The **lateral ventricle** under the floor of the corpus callosum is the cavity of the cerebral hemisphere. Cerebrospinal fluid is produced here.

The **basal ganglia or nuclei** are buried deeply in the hemisphere. They have a much more complex structure than can be shown here.

The slit-like **third ventricle** has a thalamus on both sides.

The **internal capsule** carries fibres between the brain stem and the cortex. Damage here causes loss of voluntary movement and sensation over half of the body.

The Electroencephalogram

The electrical activity of the cortex can be detected and recorded from electrodes placed at various sites over the scalp. The recorded activity, or *electroencephalogram,* represents the combined effect of many thousands of neurones acting in concert. Activity is quite irregular with the eyes open, but when the eyes are closed a regular rhythm with a frequency of about 10 cycles/sec becomes apparent. This is called the *alpha rhythm.* The rhythm changes during sleep, the exact pattern varying with the stage of sleep. The EEG is much altered by certain drugs. It is of some value in the diagnosis of diseases of the brain.

3.4. The brain stem

The brain stem connects the spinal cord to the cerebral hemispheres. It divides into the midbrain, pons and medulla. It consists of ascending and descending bundles of white matter, with *nuclei* of grey matter scattered through it. Some of these nuclei are the cells of origin of the cranial nerves; others are relay stations for ascending or descending fibres.

The major sensory relay station is the **thalamus**. Impulses ascending to the sensory cortex are first sorted and integrated here. Damage to the thalamus causes an unpleasant exaggeration of any stimulus and spontaneous pains over the opposite side of the body.

The **internal capsule** radiates from the cerebral peduncle carrying fibres to and from the cerebral cortex.

The **basal ganglia** (removed) are partly wrapped in, and partly encircle, the internal capsule. They are motor in function. They act as inhibiting centres, relaxing and smoothing willed movement. Damage here causes stiffness and tremor in the muscles.

The sensory fibres for touch and proprioception relay in two **nuclei** in the medulla.

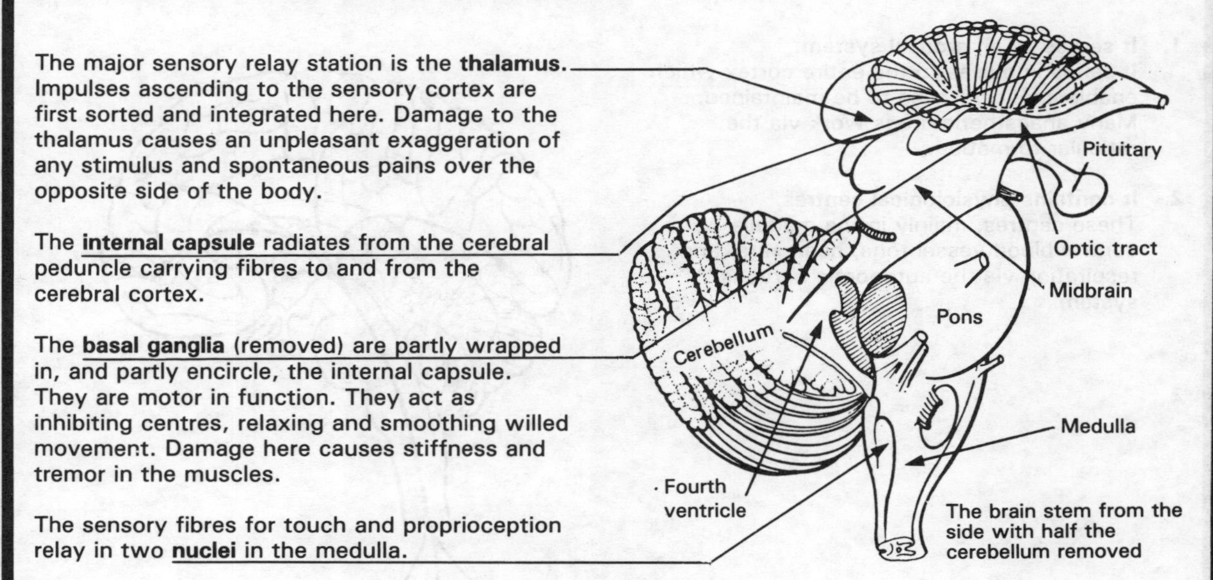

The brain stem from the side with half the cerebellum removed

These fibres, and all the other special sensory tracts, cross over to the opposite side on their way up the brain stem to the thalamus. The left hemisphere therefore appreciates sensations arising from the right side of the body.

3.5. The reticular formation

The brain stem contains throughout its length a diffuse collection of nuclei and fibres scattered in between the other structures. This is the reticular formation. It receives impulses from all ascending and descending tracts.

1. **It serves as an arousal system.**
   It sends impulses to the entire cortex which enable consciousness to be maintained. Many anaesthetic drugs work via the reticular formation.

2. **It contains physiological centres.**
   These centres, mainly in the medulla, control blood vessel tone, heart rate, and respiration via the autonomic nervous system.

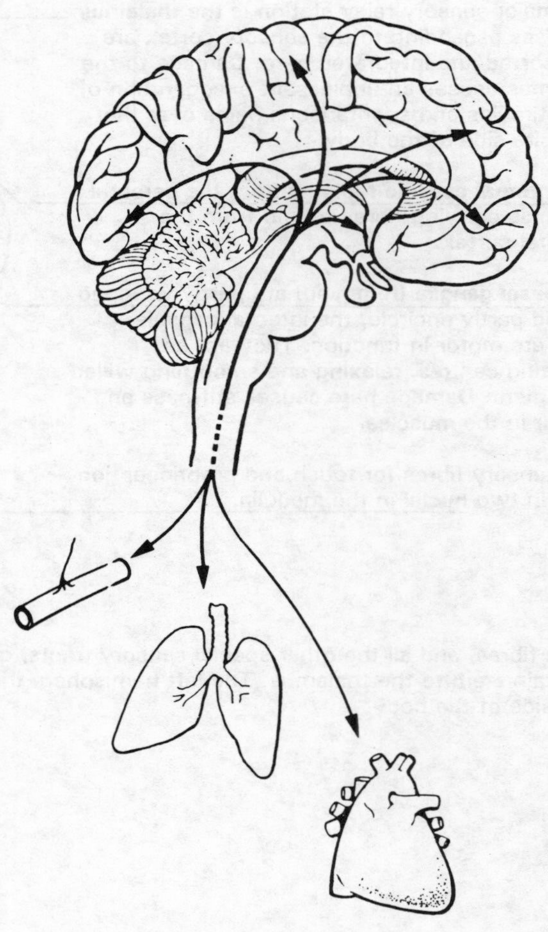

3.6. The brain viewed from below

A view of the brain from below shows the following features.

Frontal lobe

Temporal lobe

The **longitudinal fissure** separates the cerebral hemispheres.

Olfactory nerve (I).

Each **optic nerve** exchanges fibres (from nasal side of retina) to form the **optic tract** which ends in the **lateral geniculate body** (a relay station for vision) after encircling the cerebral peduncles.

The **hypothalamus** is important in the control of the pituitary gland, in electrolyte and water balance, in temperature regulation, and is the higher centre of the autonomic nervous system.

The pons is a broad encircling band connecting the two sides of the cerebellum.

The **pyramids** contain the main motor tracts from the motor cortex to the spinal cord.

They cross over in the lower medulla, so that the left hemisphere controls movement on the right side of the body.

The **cerebellum** controls posture, the underlying motor activity of the trunk and limbs on which willed movements are superimposed. It collects sensory information from the balance organs and proprioceptors and transmits motor instructions to the muscles of the trunk and limbs. It thus maintains equilibrium and smooths out and controls the movements caused by willed impulses from the cortex. It operates largely through the nerves supplying the muscle spindles. Damage to the cerebellum causes willed movement to become clumsy, exaggerated and irregular.

3.7. The cranial nerves

The cranial nerves principally supply the structures of the head and neck. There are twelve pairs in all, numbered (with Roman numerals) in the order in which they arise. Some of them are mainly sensory, some mainly motor (those shown on the page opposite) and some are mixed.

III
IV
VI
VIII
IX
X
X
XI
XII

Mainly sensory nerves

The **olfactory nerve** (I) conveys smell impulses from the upper nose by a complicated pathway to the most primitive part of the cortex, a strip above the corpus callosum.

The **optic nerve** (II), carries light impulses from the eye. It shares fibres with its partner at the **chiasma** so that the optic tract on each side conveys light impulses from the opposite visual field only.

The **trigeminal nerve** (V) is a large nerve, conveying ordinary sensation (e.g. touch, pain) from the face and mouth. It has three branches which supply forehead, cheek and lower jaw areas. It is also the motor nerve for the muscles of mastication that move the lower jaw.

The **auditory nerve** (VIII), carries sensory fibres from the cochlea (organ of hearing) and labyrinth (organ of balance).

The **glossopharyngeal nerve** (IX) is principally sensory (taste) to the back of the tongue and the pharynx — the sensory area of the gag reflex. It also carries sensory fibres from the carotid body (blood oxygenation measurement) and the carotid sinus (blood pressure measurement). It sends a small secretory (motor) supply to the parotid salivary gland (parasympathetic).

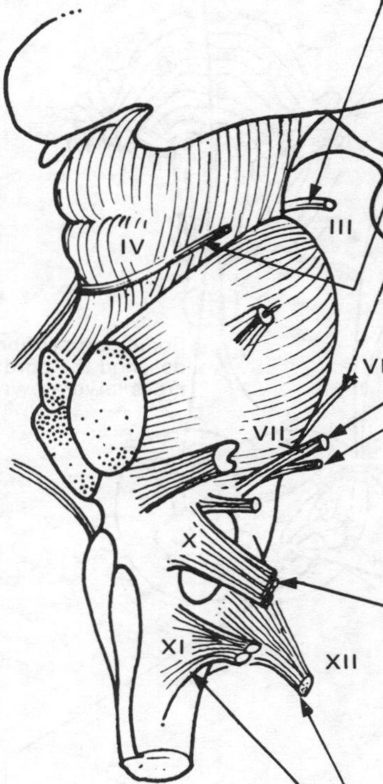

Mainly motor nerves

The **oculomotor nerve** (III) supplies four of the six eye muscles. In addition it sends parasympathetic fibres via the ciliary ganglion in the orbit to the smooth muscle of the iris and that controlling the lens.

The **trochlear nerve** (IV) supplies one eye muscle, (the superior oblique).

The **abducens nerve** (VI) supplies a single eye muscle, (the lateral rectus which abducts the eye).

The **facial nerve** (VII) supplies all the muscles of facial expression. Its path runs through the inner and middle ear.

The **nervus intermedius** (numbered usually with the facial nerve, as it fuses with it for part of its course) is a mixed nerve.

It is motor to the tear glands and the salivary glands, passing through parasympathetic ganglia on its pathway.

It is sensory for taste for the entire mouth and tongue. Its ganglion for these fibres lies on the facial nerve close to the inner ear and its main taste branch, the chorda tympani, passes through the middle ear.

The **vagus nerve** (X) is a mixed nerve which travels long distances.

It is *motor* to the muscles of the pharynx and larynx, to the smooth muscle of the stomach, small bowel and ascending colon, to the gastric glands, to the smooth muscle in the bronchial tree, and it gives an inhibitory supply to the heart.

It is *sensory* to the lower pharynx, larynx and airways. It carries stretch receptor impulses from lung tissue and blood pressure receptor impulses from the aorta.

The **accessory nerve** (XI) supplies two large neck muscles, the sternomastoid and trapezius.

The **hypoglossal nerve** (XII), supplies all the muscle bulk of the tongue and the muscles which move it.

3.8. Motor pathways

Willed movement is initiated by the **motor cortex.** The impulses travel in the pyramidal tract and descend to the motor neurone in the anterior horn.

The **pyramidal tract** crosses over in the brain stem.

The **basal ganglia** modify the effect of the cortical impulses mainly by selectively inhibiting the motor neurone, smoothing out movement. These inhibitory impulses travel down the **extrapyramidal tract.**

The **cerebellum** sends impulses along fibres to the motor neurone. These provide a background of muscle tone, to maintain posture and regulate the actions of the various muscle groups involved.

Stretch receptors in the muscle, and from tendons, joints, neighbouring muscles and opposite limbs, send fibres to the motor neurone.

Some cortical fibres descend the cord and cross lower down

Final common pathway

All of these impulses, and many others (for example from the eye, organs of balance, other parts of the cortex and brain stem) act on the **motor neurone** which is the final common pathway of their effects. The 'firing' of the neurone, and the frequency with which it 'fires', depend on the sum of all this activity.

3.9. Sensory pathways

There are two major direct pathways for sensory impulses ascending to the brain.

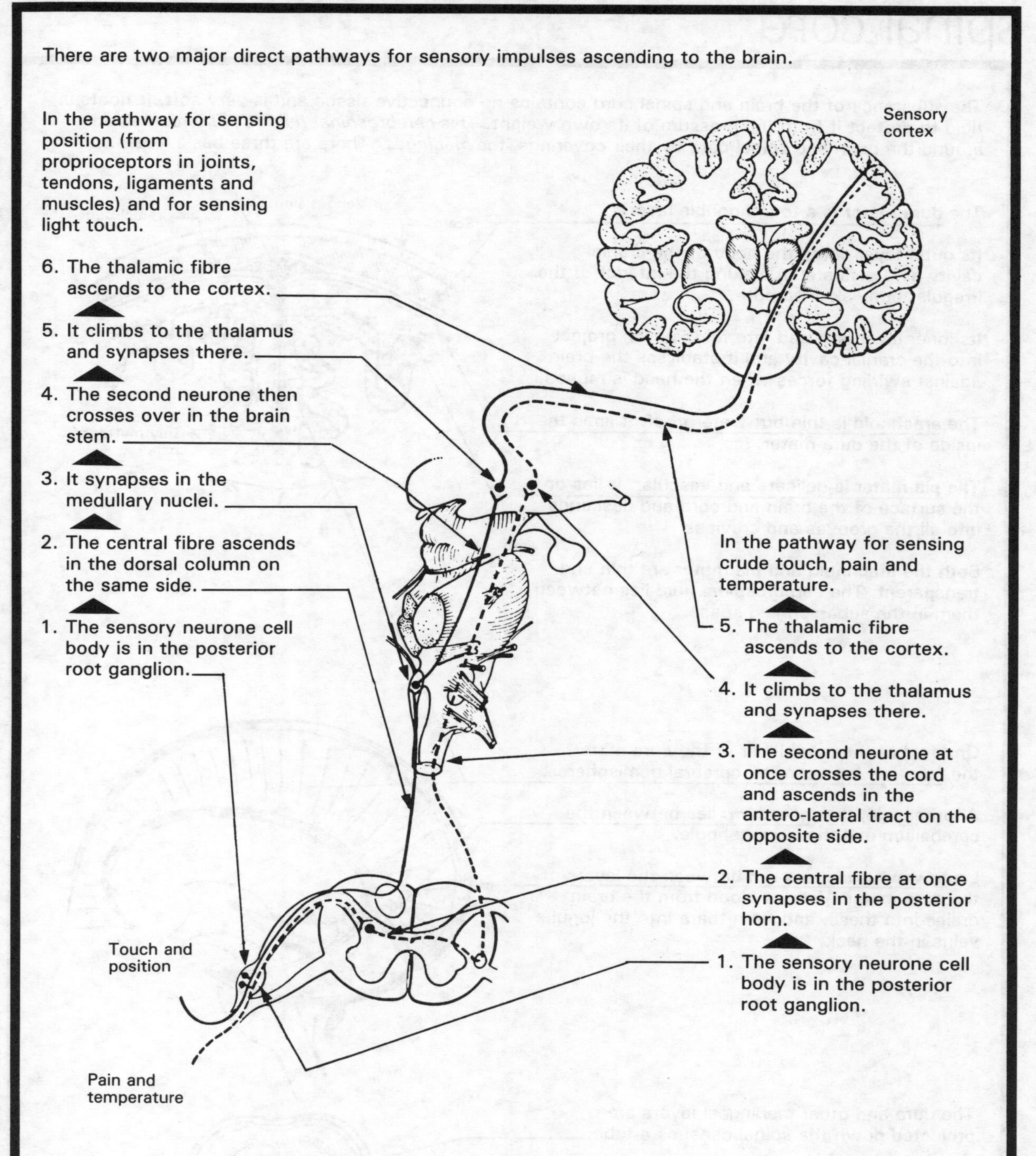

All sensory impulses from the right side of the body reach the left cortex, and vice versa.

The significance of these two different pathways is that the spinal cord may be damaged so that one set of sensations is interfered with, and yet the other set remains intact; the Brown-Sequard syndrome. Consequently afflictions of the spinal cord may result in either a limb having full motor capability but, being unable to feel pain, vulnerable to constant injury, or in a limb having limited motor capability but able to feel pain.

3.10. The coverings of the brain and spinal cord

The substance of the brain and spinal cord contains no connective tissue and is very soft. It floats in fluid to protect it from the pressure of its own weight. This *cerebrospinal fluid (CSF)* is contained around the brain and spinal cord by their coverings, the *meninges.* There are three basic layers.

The **dura mater** is a tough double layer.

Its outer layer coats the bone of the cranial cavity as a periosteum making the interior of the irregular bony skull smooth.

Its inner layer is raised into folds which project into the cranial cavity and it stabilises the brain against swirling forces when the head is moved.

The **arachnoid** is thin but waterproof. It lines the inside of the dura mater.

The **pia mater** is delicate and vascular. It lies on the surface of the brain and cord and descends into all the grooves and crevices.

Both the arachnoid and pia mater are thin and transparent. The cerebrospinal fluid lies between them in the subarachnoid space.

Venous sinus
Scalp
Cranial bone
Grey matter of cortex

One fold of the inner layer of the dura mater, the **falx,** lies between the cerebral hemispheres.

Another fold, the **tentorium,** lies between the cerebellum and the occipital pole.

Large **venous sinuses** run between the layers of the dura mater. Venous blood from the brain drains into these, and from there into the jugular veins in the neck.

The dura and other meningeal layers are projected down the spinal canal as a tube.

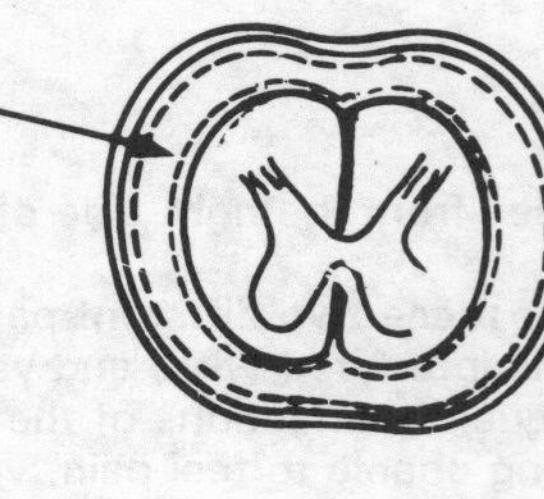

The dural tube continues to the lower end of the sacrum, although the spinal cord ends at the second lumbar disc. The lower part of the tube therefore is usually reserved for nerves emerging from the spinal canal between the vertebrae. These float in cerebrospinal fluid.

The cerebrospinal fluid (CSF) is a crystal-clear fluid which resembles blood plasma in composition, but with a much lower protein content. Normally it contains very few cells. Its volume is about 140 ml. It is present around the entire brain (filling the *ventricles* and *subarachnoid space*) and spinal cord.

Cerebrospinal fluid is produced in the **lateral ventricles,** the cavities within the hemispheres, by a strip of cauliflower-like tissue containing a mesh of blood vessels called the *choroid plexus.*

It flows from the lateral ventricles into the slit-like **third ventricle,** and then into the fourth ventricle. It escapes from the **fourth ventricle** into the subarachnoid space through three small holes in the roof of the ventricle. The fluid then circulates over the whole surface of the brain.

It is absorbed back into the bloodstream at the **arachnoid granulations.** These are hollow leaf-like structures which project into the dural venous sinuses from the subarachnoid space. The major sinus, the *saggital sinus,* runs over the brain in the mid-line in the falx.

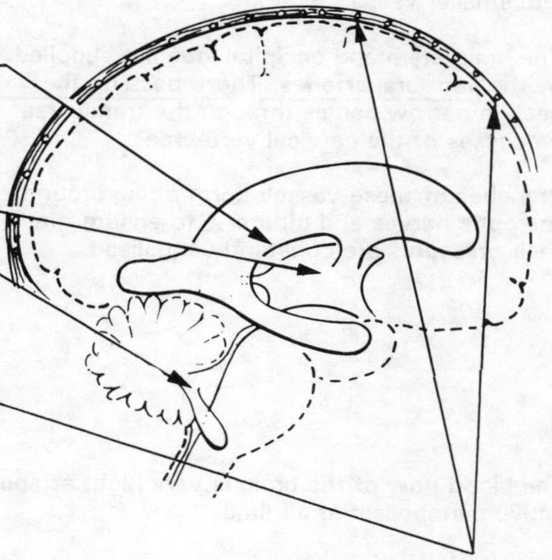

Obstruction of the free flow of cerebrospinal fluid can occur (i) within the ventricles, (ii) at narrow parts of the subarachnoid space (e.g. the foramen magnum, or at the opening in the tentorium), (iii) by clotting of blood or obstruction in the major venous sinuses. Such obstruction leads to *hydrocephalus* — an excessive accumulation of CSF. In babies, where obstruction may arise during early development, the soft skull bones yield and there is progressive enlargement of the head. In adults, the ventricles may enlarge, and there is gradual atrophy and thinning of the brain tissue. If acute, hydrocephalus can cause death in a matter of days.

3.11. The blood supply of the brain

The hemispheres are supplied by the **internal carotid arteries** which pass up on each side of the neck and enter the base of the skull alongside the pituitary gland, before they branch into smaller vessels.

The brain stem and occipital lobe are supplied by the **vertebral arteries.** These pass up the neck in narrow canals through the transverse processes of the cervical vertebrae.

Branches of these vessels form a ring around the optic nerves and pituitary, to ensure that their pressures are constantly equalised.

The blood flow of the brain is very high, at about 750 ml per minute, as the neurones have a very active metabolism at all times.

3.12. The functions of the brain

There is still a great deal of mystery about the functioning of the brain. The significance of such anatomical features as the double cortex (dividing the corpus callosum causes little disability), the crossing over of the major nerve tracts, and the segregation into grey and white matter is unknown.

More importantly, the true nature of such functions as memory, reason, and intellectual analysis is just not understood. These functions appear to be carried out diffusely by the cortex as a whole. Large parts of the cortex may be damaged without destroying long-established memory. Even such basic aspects of reason as the encoding of meaningful messages into the sounds of speech and their subsequent interpretation by the cortex of a second person is beyond the ability of even the most elaborate computer system which can yet be envisaged. It apparently involves principles that are still to be discovered.

TEST THREE

1. Indicate which of the names in the list below refer to the parts of the developing brain labelled on the diagram alongside, by placing the appropriate letters in the brackets.

1. Midbrain. ( )
2. Medulla. ( )
3. Corpus callosum. ( )
4. Cerebral hemisphere. ( )
5. Cerebellum. ( )

A
C
B
D
E

2. (a) What is the name of the coat of grey matter over the hemispheres?

____

(b) What is the cavity of the hemisphere called?

____

3. What is the reticular formation, and what are its functions?

____

4. Indicate which of the names in the list below refer to the parts of the brain labelled on the diagram alongside, by placing the appropriate letters in the brackets.

1. Frontal lobe. ( )
2. Occipital lobe. ( )
3. Temporal lobe. ( )
4. Sensory cortex. ( )
5. Parietal lobe. ( )
6. Motor cortex. ( )

B
C
A
D
E
F

5. Which lobe is concerned with the appreciation of the weight, texture or identity of an object?

____

6. Which of the statements on the right below apply to the nerves on the left?

(i) The trigeminal nerve (V). (a) Conveys light impulses.

(ii) The facial nerve (VII). (b) Conveys ordinary sensation from the face.

(iii) The optic nerve (II). (c) Is the motor nerve of the facial muscles.

ANSWERS TO TEST THREE

1. 1. Midbrain. (B)

2. Medulla. (E)

3. Corpus callosum. (C)

4. Cerebral hemisphere. (A)

5. Cerebellum. (D)

2. (a) The cortex.

(b) The lateral ventricle.

3. The reticular formation is a collection of neurones in the brain stem, through which impulses pass in both directions between the brain and spinal cord. The reticular formation is involved in the maintenance of consciousness, in emotional states, and in the control of blood vessel tone, heart rate and respiration.

4. 1. Frontal lobe. (A)

2. Occipital lobe. (E)

3. Temporal lobe. (F)

4. Sensory cortex. (C)

5. Parietal lobe. (D)

6. Motor cortex. (B)

5. The parietal lobe.

6. (i) The trigeminal nerve (V) (b) conveys ordinary sensation from the face.

(ii) The facial nerve (VII) (c) is the motor nerve of the facial muscles.

(iii) The optic nerve (II) (a) conveys light impulses.

4. The autonomic nervous system

4.1. Introduction

The autonomic (involuntary) nervous system carries impulses to smooth muscle, cardiac muscle, and glands.

The main structural difference between the autonomic nervous system and the voluntary nervous system is in the arrangement of the effector neurones:

— in the somatic nervous system, one **neurone** conducts an impulse from the CNS to the effector (e.g. skeletal muscle)

— in the autonomic nervous system there are two neurones:

— a **pre-ganglionic neurone** which synapses in a ganglion
— with one or more **post-ganglionic neurones** which take impulses to the effectors (e.g. glands, blood vessels or smooth muscle in the walls of the viscera).

The fact that one pre-ganglionic neurone may synapse with several post-ganglionic neurones accounts for the diffuseness of response possible in the autonomic nervous system.

The activity of the autonomic nervous system is controlled and co-ordinated by the hypothalamus of the brain.

There are two main divisions of the autonomic nervous system:

(1) the *parasympathetic* whose pre-ganglionic neurones leave the CNS via cranial nerves III, VII, IX and X and via the **sacral segments** of the spinal cord.

The transmitter substance at the synapses of the parasympathetic nervous system is *acetylcholine.*

The transmitter substance at the junction between the post-ganglionic neurones and the effectors is also *acetylcholine.*

This division is also called the *cholinergic* division of the autonomic nervous system.

(2) the *sympathetic* whose pre-ganglionic neurones leave the CNS via the thoracic and first two lumbar segments of the spinal cord.

The transmitter substance at the synapses of the sympathetic nervous system is *acetylcholine.*

The transmitter substance at the junction between the post-ganglionic neurones and the effectors is *noradrenalin.*

This division is also called the *adrenergic* division of the autonomic nervous system.

4.2. The parasympathetic nervous system

In this system the pre-ganglionic fibres are long. In the head and neck they synapse in ganglia close to the base of the skull. In the rest of the body and ganglia are in the wall of the organ supplied.

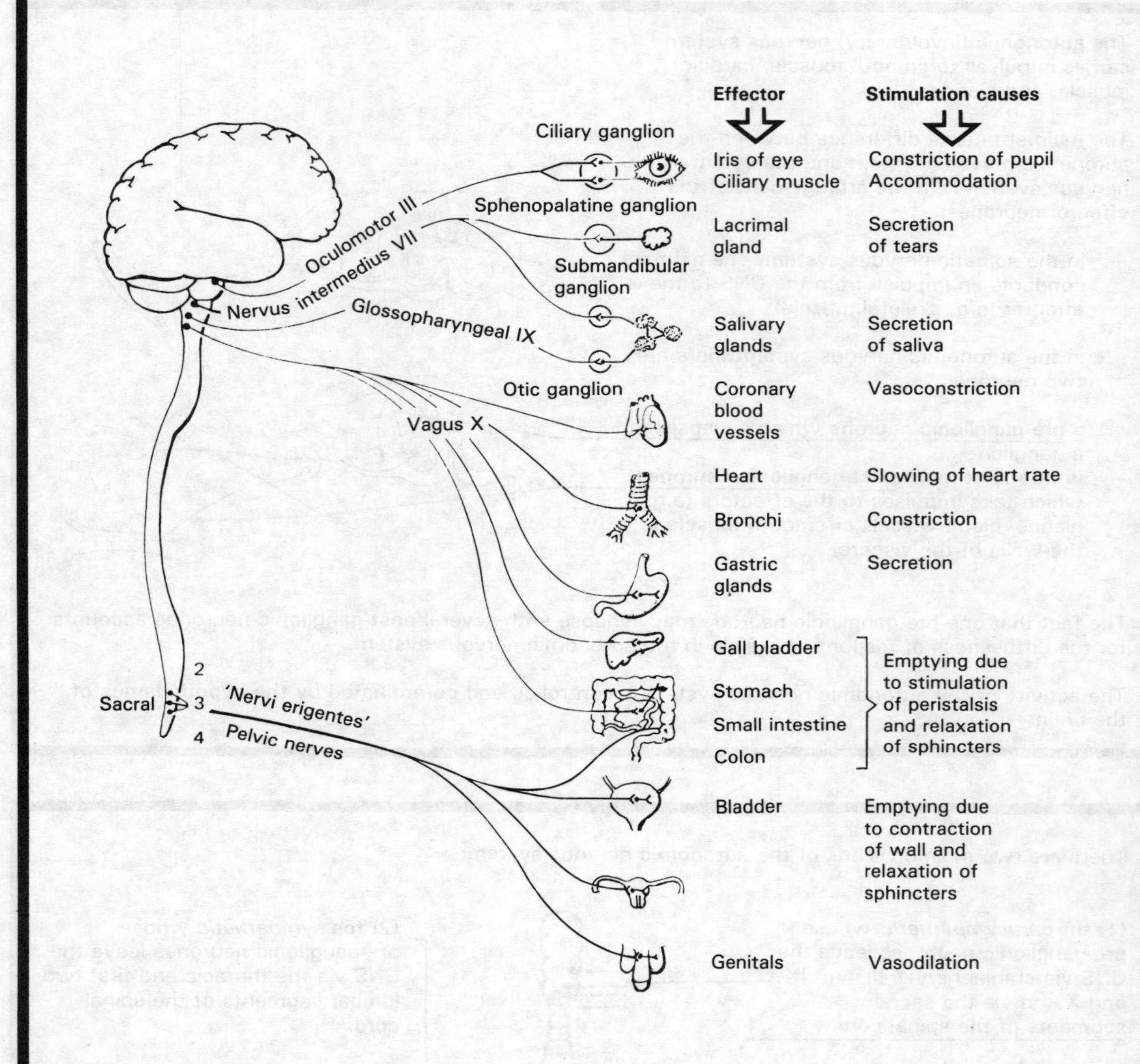

The cholinergic division is important in normal control of digestion, excretion, and metabolism; it balances the action of the adrenergic divisions, which is more of an *emergency* mechanism.

4.3. The sympathetic nervous system

The effects of the sympathetic nervous system tend to be opposite to those of the parasympathetic. Its most important effect is causing constriction of blood vessels thus controlling blood flow. Most of the sympathetic nervous supply to any organ goes to its blood vessels.

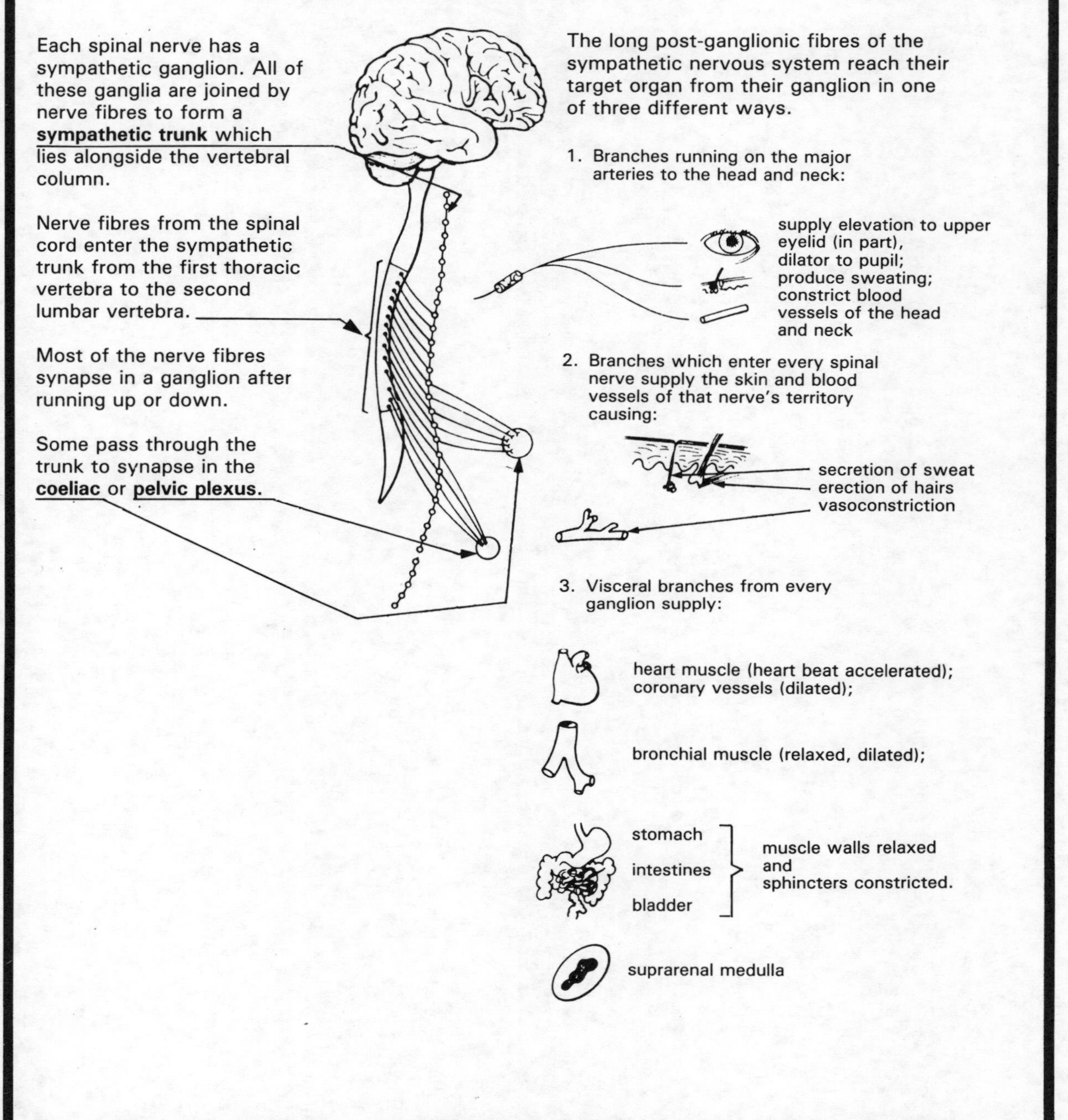

The effects of sympathetic nervous stimulation are generally helpful in emergency situations ('fight or flight') since they cause blood to be diverted to useful areas.

There is a sympathetic vasodilator supply to limb muscles. If these nerves are stimulated by a fright without activity, the blood pressure will fall and the fall may be sufficient to cause fainting.

TEST FOUR

1. **Name two of the cranial nerves which carry parasympathetic fibres.**

   ____

2. (a) **What is the chemical transmitter at the autonomic ganglia?**

   ____

   (b) **What is the chemical transmitter at the junction of the post-ganglionic fibres and the effectors of the sympathetic nervous system?**

   ____

3. **What effect does stimulation of the vagus nerve have on the heart?**

   ____

4. **Complete the following:**
   In the sympathetic division of the ________ nervous system (which is also called the division) the pre-ganglionic neurones leave the ________ via the ________ and first two segments of the ________. At the synapses ________ is the chemical transmitter, while at the junction between post-ganglionic neurones and the effectors, the chemical transmitter is ________.

5. **In general, when is overall sympathetic activity useful?**

   ____

ANSWERS TO TEST FOUR

1. Two of the following:
The oculomotor nerve.
The glossopharyngeal nerve.
The vagus nerve.

2. (a) Acetylcholine.

(b) Noradrenalin.

3. It slows the rate.

4. In the sympathetic division of the *autonomic* nervous system (which is also called the *adrenergic* division) the pre-ganglionic neurones leave the *CNS* via the *thoracic* and first two *lumbar* segments of the *spinal cord.* At the synapses *acetylcholine* is the chemical transmitter, while at the junction between post-ganglionic neurone and the effectors, the chemical tranmsitter is *noradrenalin.*

5. In an emergency situation ('fight or flight').

POST TEST

1. The central nervous system, the peripheral nervous system, the voluntary nervous system, the autonomic nervous system: what do these terms have in common and how do they differ?

2. Which of the parts of the action potential curve shown alongside are associated with:
(i) entry of sodium ions into the fibre?

(ii) escape of potassium ions from the fibre?

0
B
A

3. Which of the regions labelled on the diagram alongside are associated with motor neurones or fibres, and which are associated with sensory neurones or fibres?

B
C
A
D
E

4. Which of the statements on the right apply to the nerves listed on the left?

(i)	The radial nerve.	(a)	Supplies the extensors of the arm.
(ii)	The thoracic nerve.	(b)	Supplies the quadriceps of the thigh.
(iii)	The femoral nerve.	(c)	Supplies the large calf muscles.
(iv)	The tibial nerve.	(d)	Supplies the muscles of the abdominal wall.

5. Indicate which of the names in the list below refer to the parts of the developing brain labelled on the diagrams alongside by placing the appropriate letters in the brackets.

1. **Pons.** ( )
2. **Cortex.** ( )
3. **Corpus callosum.** ( )
4. **Thalamus.** ( )
5. **Hypothalamus.** ( )
6. **Basal ganglia.** ( )

A
B
C D E F

6. Give two functions of the hypothalamus.

POST TEST

7. Complete the following:
Thoughts are formulated into words in the ________ cortex which for people who are________ , lies in the________hemisphere of the brain. The________ lobe is concerned with co-ordinating and understanding what is seen, smelt, touched, tasted or heard, while the ________ cortex is concerned with recognition of identity and the assessment of time, weight, distance, etc.

8. Indicate which of the names in the list below apply to the parts of the brain labelled on the diagram alongside, by placing the appropriate letters in the brackets.

1. Pituitary. ( )
2. Cerebral peduncle. ( )
3. Pons. ( )
4. Third ventricle. ( )
5. Corpus callosum. ( )

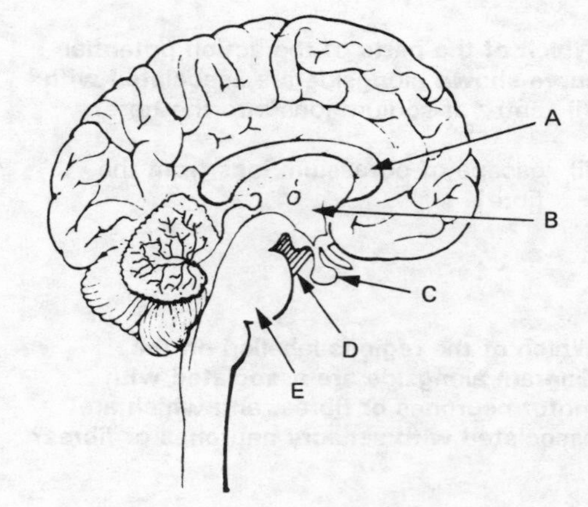

9. Place the following parts of the pyramidal (motor) tract in descending order.

(a) Internal capsule.

(b) Cerebral peduncle.

(c) Motor cortex.

(d) Anterior horn.

10. Which of the statements on the right apply to the cranial nerves listed on the left?

(i) The olfactory nerve (I).	**(a) carries smell impulses.**
(ii) The auditory nerve (VIII).	**(b) supplies the muscles of the tongue.**
(iii) The oculomotor nerve (III).	**(c) carries impulses from organs of balance.**
(iv) The hypoglossal nerve (XII).	**(d) supplies most of the eye muscles.**

11. (a) Name the three layers of the meninges.

____

(b) Which of these is the toughest layer?

____

12. Answer the following questions about cerebrospinal fluid.

(a) Where is CSF found?
(b) What is its appearance and composition?
(c) Where is it produced?
(d) How does it gain access to the brain surface?
(e) How is it absorbed back into the bloodstream?

ANSWERS TO POST TEST

1. The terms in question are used to describe the nervous system, but while the division into a central and a peripheral nervous system is purely one of convenience, the distinction between the voluntary and the autonomous nervous systems relates to clear differences in function.

2. (i) A.

(ii) B.

3. A Sensory.

B Sensory.

C Motor.

D Motor.

E Motor.

4.

(i)	The radial nerve	(a)	supplies the extensors of the arm.
(ii)	The thoracic nerve	(d)	supplies the muscles of the abdominal wall.
(iii)	The femoral nerve	(b)	supplies the quadriceps of the thigh.
(iv)	The tibial nerve	(c)	supplies the large calf muscles.

5.

1.	Pons.	(B)
2.	Cortex.	(C)
3.	Corpus callosum.	(A)
4.	Thalamus.	(E)
5.	Hypothalamus.	(D)
6.	Basal ganglia.	(F)

6. Two of the following:

Control of the pituitary gland.
Regulation of water and electrolyte balance.
Regulation of body temperature.
Serving as the higher centre of the autonomic nervous system.

ANSWERS TO POST TEST

7. Thoughts are formulated into words in the *speech* cortex which for people who are *right-handed,* lies in the *left* hemisphere of the brain. The *temporal* lobe is concerned with co-ordinating and understanding what is seen, smelt, touched, tasted or heard, while the *parietal* cortex is concerned with recognition of identity and the assessment of time, weight, distance, etc.

8.

1.	Pituitary.	(C)
2.	Cerebral peduncle.	(D)
3.	Pons.	(E)
4.	Third ventricle.	(B)
5.	Corpus callosum.	(A)

9. (c) Motor cortex.

(a) Internal capsule.

(b) Cerebral peduncle.

(d) Anterior horn.

10.

(i) The olfactory nerve (I)	(a)	carries smell impulses.
(ii) The auditory nerve (VIII)	(c)	carries impulses from organs of balance.
(iii) The oculomotor nerve (III)	(d)	supplies most of the eye muscles.
(iv) The hypoglossal nerve (XII)	(b)	supplies the muscles of the tongue.

11. (a) Dura mater, arachnoid, pia mater.

(b) Dura mater.

12. (a) CSF surrounds the entire brain and spinal cord.
(b) It is a clear fuid, resembling blood plasma in composition but with a lower protein content.
(c) It is produced in the lateral ventricles by the choroid plexus.
(d) From the lateral ventricles CSF flows into the third and then the fourth ventricles. The fluid then enters the subarachnoid space and circulates over the whole brain surface.
(e) CSF is absorbed back into the bloodstream via the arachnoid granulations.